Springer Proceedings in Mathematics & Statistics

Volume 30

For further volumes:
http://www.springer.com/series/10533

Springer Proceedings in Mathematics & Statistics

This book series features volumes composed of select contributions from workshops and conferences in all areas of current research in mathematics and statistics, including OR and optimization. In addition to an overall evaluation of the interest, scientific quality, and timeliness of each proposal at the hands of the publisher, individual contributions are all refereed to the high quality standards of leading journals in the field. Thus, this series provides the research community with well-edited, authoritative reports on developments in the most exciting areas of mathematical and statistical research today.

Pedro Latorre Carmona • J. Salvador Sánchez
Ana L.N. Fred
Editors

Mathematical Methodologies in Pattern Recognition and Machine Learning

Contributions from the International Conference on Pattern Recognition Applications and Methods, 2012

Springer

Editors
Pedro Latorre Carmona
Depto. Lenguajes y Sistemas Informáticos
Jaume I University
Castellón de la Plana, Spain

J. Salvador Sánchez
Depto. Lenguajes y Sistemas Informáticos
Jaume I University
Castellón de la Plana, Spain

Ana L.N. Fred
Technical University of Lisbon
Lisbon, Portugal

ISSN 2194-1009 ISSN 2194-1017 (electronic)
ISBN 978-1-4614-5075-7 ISBN 978-1-4614-5076-4 (eBook)
DOI 10.1007/978-1-4614-5076-4
Springer New York Heidelberg Dordrecht London

Library of Congress Control Number: 2012952083

Mathematics Subject Classification (2010): 93-XX, 90-99, 68-06

Printed on acid-free paper

Springer is part of Springer Science+Business Media (www.springer.com)

Preface

This volume features key contributions from the International Conference on Pattern Recognition Applications and Methods (ICPRAM 2012) held in Vilamoura, Algarve, Portugal from February 6 to 8, 2012.

ICPRAM was sponsored by the Institute for Systems and Technologies of Information Control and Communication (INSTICC) and held in cooperation with the Association for the Advancement of Artificial Intelligence (AAAI) and Pattern Analysis, Statistical Modelling and Computational Learning (PASCAL2). It was technically co-sponsored by IEEE Signal Processing Society, Machine Learning for Signal Processing (MLSP) Technical Committee of IEEE, AERFAI (Asociación Española de Reconocimiento de Formas y Análisis de Imagen) and APRP (Associação Portuguesa de Reconhecimento de Padrões).

ICPRAM received 259 paper submissions from 46 countries in all continents. To evaluate each submission, a double-blind paper review was performed by the Program Committee, whose members are highly qualified researchers in ICPRAM topic areas. Based on the classifications provided, only 115 papers were selected for oral presentation (61 full papers and 54 short papers) and 32 papers were selected for poster presentation. The full paper acceptance ratio was 24 %, and the total oral acceptance ratio (including full papers and short papers) was 44 %. These strict acceptance ratios show the intention to preserve a high quality forum which we expect to develop further next year.

The conference provided a major point of collaboration between researchers, engineers and practitioners in the areas of Pattern Recognition, both from theoretical and applied perspectives. Contributions described applications of pattern recognition techniques to real-world problems, interdisciplinary research, and experimental and theoretical studies.

This book will be suitable for scientists and researchers in optimization, numerical methods, computer science, statistics and for differential geometers and mathematical physicists.

Castellón de la Plana, Spain — Pedro Latorre Carmona
Castellón de la Plana, Spain — J. Salvador Sánchez
Lisbon, Portugal — Ana L.N. Fred

Contents

On the Expressivity of Alignment-Based Distance and Similarity Measures on Sequences and Trees in Inducing Orderings

Martin Emms and Hector-Hugo Franco-Penya

Abstract Both 'distance' and 'similarity' measures have been proposed for the comparison of sequences and for the comparison of trees, based on scoring mappings. For a given alphabet of node-labels, the measures are parameterised by a table giving label-dependent values for swaps, deletions and insertions. The paper addresses the question whether an ordering by a 'distance' measure, with some parameter setting, can be also expressed by a 'similarity' measure, with some other parameter setting, and vice versa. Ordering of three kinds is considered: alignment-orderings, for fixed source S and target T, neighbour-orderings, where for a fixed S, varying candidate neighbours T_i are ranked, and pair-orderings, where for varying S_i, and varying T_j, the pairings $\langle S_i, T_j \rangle$ are ranked. We show that (1) any alignment-ordering expressed by 'distance' setting be re-expressed by a 'similarity' setting, and vice versa; (2) any neigbour-ordering and pair-ordering expressed by a 'distance' setting be re-expressed by a 'similarity' setting; (3) there are neighbour-orderings and pair-orderings expressed by a 'similarity' setting which *cannot* be expressed by a 'similarity' setting. A consequence of this is that there are categorisation and hierarchical clustering outcomes which can be achieved via similarity but not via distance.

Keyword Similarity distance tree sequence

M. Emms (✉) • H.-H. Franco-Penya
School of Computer Science and Statistics, Trinity College, Dublin, Ireland
e-mail: Martin.Emms@tcd.ie; francoph@tcd.ie

P. Latorre Carmona et al. (eds.), *Mathematical Methodologies in Pattern Recognition and Machine Learning*, Springer Proceedings in Mathematics & Statistics 30,
DOI 10.1007/978-1-4614-5076-4_1,

1 Tree Distance and Similarity

In many pattern-recognition scenarios the data either takes the form of, or can be encoded as, sequences or trees. Accordingly, there has been much work on the definition, implementation and deployment of measures for the comparison of sequences and for the comparison of trees.

These measures are sometimes described as 'distances' and sometimes as 'similarities'. We are concerned in what follows in first distinguishing between these, and then with the question whether orderings induced by a 'distance' measure can be dualized by a 'similarity' measure, and vice versa. To an extent this can be seen as providing for sequences and trees a counterpart to the kind of analysis that has been applied to set and vector comparison measures [1, 10, 11].

From statements such as the following

> *To compare RNA structures, we need a score system, or alternatively a distance, which measures the similarity (or the difference) between the structures. These two versions of the problem score and distance are equivalent.* [7]

which are not uncommon in the literature it would be easy to gain the impression that similarity and distance (on sequences and trees) are straightforwardly interchangeable notions. In Sect. 1.1 several distinct kinds of equivalence are defined. Sections 2, 3.1 and 3.2 then show that while some kinds of equivalence hold, others do not.

To begin we need to clarify what we will mean by 'distance' and 'similarity' on sequences and trees. Because sequences can be encoded as vertical trees it suffices to give definitions for trees. Tai [16] first proposed a tree-distance measure, based on scoring a mapping. Where S and T are ordered, labelled trees, a *Tai* mapping $\sigma : S \mapsto T$ is a *partial, 1-to-1* function from the nodes of S into the nodes of T, which respects *left-to-right order* and *ancestry*.[1] Figure 1a shows an example Tai-mapping. To define how the score is assigned, it is convenient to identify three sets related to the mapping:

$$\mathcal{M}: \text{ the pairs } (i,j) \in \sigma \qquad \begin{cases} \mathcal{D}: \text{ those } i \in S \text{ s.t. } \forall j \in T, (i,j) \notin \sigma \\ \mathcal{I}: \text{ those } j \in T \text{ s.t. } \forall i \in S, (i,j) \notin \sigma \end{cases}$$

The scoring is based on the these 'match', 'deletion' and 'insertion' sets in a label-dependent way, and it is conventional to specify the costs with a table—call it C^{Δ}—indexed by $\{\lambda\} \cup \Sigma$, where Σ is the alphabet of labels. Where $(.)^{\gamma}$ gives the label of a node, the table assigns 'costs' to $\mathcal{M}$, $\mathcal{D}$ and $\mathcal{I}$ according to[2]:

[1] So if (i,j) and (i',j') are in the mapping then (T1) $left(i,i')$ iff $left(j,j')$ and (T2) $anc(i,i')$ iff $anc(j,j')$.

[2] Note in this general setting even a pairing of two nodes with identical labels can in principal make a nonzero cost contribution.

a

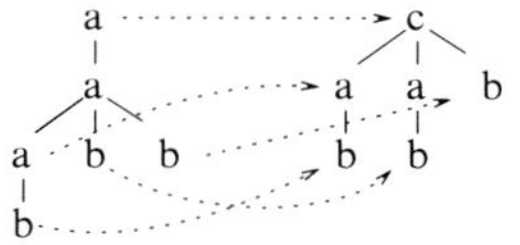

b *With* $C^{\Delta}(x,\lambda) = C^{\Delta}(\lambda,x) = 1$, $C^{\Delta}(x,x) = 0$, $C^{\Delta}(x,y) = 1$ *for* $x \neq y$, *the alignment has score* $\Delta(\sigma) = 3$ *and this is minimal for the given* C^{Δ}

c *With* $C^{\Theta}(x,\lambda) = C^{\Theta}(\lambda,x) = 0$, $C^{\Theta}(x,x) = 2$, $C^{\Delta}(x,y) = 1$ *for* $x \neq y$, *the alignment has score* $\Theta(\sigma) = 9$ *and this is maximal for the given* C^{Θ}

Fig. 1 (**a**) a Tai mapping (**b**) a 'distance' scoring (**c**) a 'similarity' scoring

$$(i,j) \in \mathcal{M},\ \text{cost} = C^{\Delta}(i^{\gamma}, j^{\gamma}) \quad \begin{cases} i \in \mathcal{D}, & \text{cost} = C^{\Delta}(i^{\gamma}, \lambda) \\ j \in \mathcal{I}, & \text{cost} = C^{\Delta}(\lambda, j^{\gamma}) \end{cases}$$

Where $\sigma : S \mapsto T$ is any mapping from S to T, define $\Delta(\sigma : S \mapsto T)$ by

Definition 1 ('distance' scoring of an alignment given C^{Δ}).

$$\Delta(\sigma : S \mapsto T) = \sum_{(i,j)\in\mathcal{M}} C^{\Delta}(i^{\gamma}, j^{\gamma}) + \sum_{i\in\mathcal{D}} C^{\Delta}(i^{\gamma}, \lambda) + \sum_{j\in\mathcal{I}} C^{\Delta}(\lambda, j^{\gamma})$$

From this costing of alignments, a 'distance' score on tree pairs is defined by minimization:

Definition 2 ('distance' scoring of a tree pair given C^{Δ}). The Tree- or Tai-distance $\Delta(S,T)$ between two trees S and T is the *minimum* value of $\Delta(\sigma : S \mapsto T)$ over possible *Tai*-mappings from S to T, relative to a chosen cost table C^{Δ}.

See Fig. 1b for an example. Sequences can be encoded as vertical trees, and on this domain of trees the tree distance coincides with a well known comparison measure on sequences, the (alphabet-weighted) string edit distance [5, 17]. The definition[3] was given in terms of costs applied to a *mapping*. There is an alternative definitional route via the notion of an *edit-script* of operations, transforming S to T. For both sequences and trees the mapping-based and script-based notions coincide [9, 16, 17] and so we omit further details of the definition via edit-scripts.

There is an efficient algorithms for $\Delta(S,T)$ [18]. While the correctness of this algorithm—ie. that it truly finds the *minimal* value of $\Delta(\sigma : S \mapsto T)$ given cost-table C^{Δ}—does not require the cost-table C^{Δ} to satisfy any particular properties, some settings of C^{Δ} clearly make little sense. The combination of deletion/insertion cost-entries which are *negative*—$C^{\Delta}(x,\lambda) < 0$, $C^{\Delta}(\lambda,y) < 0$—with swap/match cost entries which are *not negative* gives the counter-intuitive effect that a supertree[4]

---

[3]The literature contains quite a number of inequivalent notions, all referred to as 'tree distance'; in this article Definition 2 will be understood to define the term.

[4]Or a subtree.

of S is 'closer'—in the sense of having a lower Δ score—to S than S itself. This is a rationale for the following nonnegativity assumption

$$\forall x,y \in \Sigma(C^{\Delta}(x,y) \geq 0,\ \ C^{\Delta}(x,\lambda) \geq 0,\ \ C^{\Delta}(\lambda,y) \geq 0) \tag{1}$$

which is a pretty universal assumption, and from which it follows that $\Delta(S,T) \geq 0$, giving a minimum consistency with the every day notion of 'distance'. In this article we will confine attention to 'distance' Δ based on a table C^{Δ} which satisfies at least (1). The question whether anything is changed in the claims we wish to make when the cost-table is constrained more strictly than this—for example requiring satisfaction of all the conditions of a *distance-metric*—we leave to Sect. 5.

Turning now to 'similarity', rather than approach the problem of comparison by *minimizing* accumulated costs assigned to an alignment, a widely followed alternative, especially for sequence comparison, has been to *maximize* a score assigned to an alignment.

Let C^{Θ} be a 'similarity' table, again indexed by $\{\lambda\} \cup \Sigma$, where Σ is the alphabet of labels, and where $\sigma : S \mapsto T$ is any mapping from S to T, and then let $\Theta(\sigma : S \mapsto T)$ be defined by

Definition 3 ('similarity' scoring of an alignment given C^{Θ}).

$$\Theta(\sigma : S \mapsto T) = \sum_{(i,j)\in\mathcal{M}} C^{\Theta}(i^{\gamma}, j^{\gamma}) - \sum_{i\in\mathcal{D}} C^{\Theta}(i^{\gamma},\lambda) - \sum_{j\in\mathcal{I}} C^{\Theta}(\lambda, j^{\gamma})$$

From this costing of alignments, a 'similarity' score on tree pairs is defined by maximisation:

Definition 4 ('similarity' scoring of a tree pair given C^{Θ}). The Tree- or Tai-similarity $\Theta(S,T)$ between two trees S and T is the *maximum* value of $\Theta(\sigma : S \mapsto T)$ over possible Tai-mappings from S to T, relative to a chosen cost table C^{Θ}

See Fig. 1c for an example. $\Theta(S,T)$ can be computed via a simple modification of the algorithm of [18]. Again on the domain of vertical trees this coincides with a well-known approach to sequence comparison, the (alphabet-weighted) string similarity [5, 13].

As with Δ, while the correctness of the algorithm for Θ is not dependent on any assumptions about the cost-table C^{Θ}, some settings of C^{Θ} make little sense. Given the formulation in (3), which *subtracts* the contribution from deletions and insertions, a setting where deletion/insertion cost entries are negative—$C^{\Theta}(x,\lambda) < 0$, $C^{\Theta}(\lambda,x) < 0$—gives the counter-intuitive effect that a supertree of S would be more 'similar'—in the sense of higher Θ score—to S than S itself. This gives a rationale for the nearly universal assumption of nonnegative deletion/insertions entries in C^{Θ}:

$$\forall x,y \in \Sigma(C^{\Theta}(x,\lambda) \geq 0,\ \ C^{\Theta}(\lambda,y) \geq 0). \tag{2}$$

In what follows we will confine attention always to 'similarity' Θ based on a table C^{Θ} satisfying (2).[5] For the C^{Θ}-entries which are not deletions or insertions, it is quite common in biological sequence comparison to have both positive and negative entries. The question whether anything is changed in the claims we wish to make should C^{Θ} be more strictly constrained we leave to Sect. 5.

To reiterate, for the purposes of this discussion a tree 'distance' measure will imply a cost-table C^{Δ}, satisfying (1), used in accordance with Definition 1 and Definition 2 to score alignments and tree pairs. A tree 'similarity' measure measure will imply a cost-table C^{Θ}, satisfying (2), used in accordance with Definition 3 and Definition 4 to score alignments and tree pairs. This is sufficient to distinguish the 'distance' approach from the 'similarity' approach in an intuitive way without commiting to any further axioms.

Also note the following concerning the relationship of these notions to other notions in the literature. Often the contribution to the score due to the deleted and inserted nodes is formulated in terms of *gap-penalty* functions, which apply to a *sequence* of consecutive deletions or insertions. Our definitions of Δ and Θ effectively coincide with the simplest possible case of such functions, where the value on a sequence is a sum on the individual parts. Also our definitions of Δ and Θ concern only what are sometimes termed 'global' alignments, in contrast to 'local' variants, such as the 'local' similarity, popular in biological sequence comparison, which seeks to maximise a 'global' score on pairs of subsequences [5].

1.1 Ordering Expressivity for Tai Distance and Similarity

Given a 'distance' Δ scoring of alignments, it can be set to work to induce orderings of at least three different kinds of entities.

Alignment ordering. Given fixed S, and fixed T, rank the possible *alignments* $\sigma : S \mapsto T$ by $\Delta(\sigma : S \mapsto T)$

Neighbour ordering. Given fixed S, and varying candidate neighbours T_i, rank the *neighbours* T_i by $\Delta(S, T_i)$—typically used in k-NN classification.

Pair ordering. Given varying S_i, and varying T_j, rank the *pairings* $\langle S_i, T_j \rangle$ by $\Delta(S_i, T_j)$—typically used in hierarchical clustering.

Similarly a 'similarity' Θ scoring of alignments induces orderings of the above kinds of entities. Comparing these orderings motivates the following definition

Definition 5 (A-, N- and P-dual). When the alignment orderings induced by a choice of C^{Δ} and by a choice C^{Θ} are the *reverse* of each other, we will say that

[5] While Definition 3 formulates Θ with deletion/insertion contributions subtracted, as is often done [13, 15], an alternative formulation has these treated additively [5]. With the additive formulation, the same consideration suggests making deletion/insertions non-positive.

C^Θ is a **A-dual** of C^Δ. Similarly we will say we have an **N-dual** when neighbour ordering is reversed, and a **P-dual** where pair-ordering is reversed.

For example, the following are A-duals in this sense (proven in Sect. 2):

Example 1. Δ with $C^\Delta(x,\lambda) = 1$ $C^\Delta(x,x) = 0$ $C^\Delta(x,y) = 1$, otherwise
Θ with $C^\Theta(x,\lambda) = 0$ $C^\Theta(x,x) = 2$ $C^\Theta(x,y) = 1$, otherwise

Example 2. Δ with $C^\Delta(x,\lambda) = 0.5$ $C^\Delta(x,x) = 0$ $C^\Delta(x,y) = 0.5$, otherwise
Θ with $C^\Theta(x,\lambda) = 0$ $C^\Theta(x,x) = 1$ $C^\Theta(x,y) = 0.5$, otherwise

If for *every* choice of C^Δ, there is a choice of C^Θ which is a A-dual, and vice versa, there is a clear sense in which distance and similarity are indistinguishable in terms of the alignment orderings they are capable of expressing. The natural question that presents itself is then whether this is so, along with other related questions concerning N- and P-duals. More precisely there are the following *order-expressibility conjectures*

$$\textit{A-duality}\begin{cases} (i)\ \forall C^\Delta \exists C^\Theta (C^\Delta \text{ and } C^\Theta \text{ are A-duals}) \\ (ii)\ \forall C^\Theta \exists C^\Delta (C^\Delta \text{ and } C^\Theta \text{ are A-duals}) \end{cases}$$

$$\textit{N-duality}\begin{cases} (i)\ \forall C^\Delta \exists C^\Theta (C^\Delta \text{ and } C^\Theta \text{ are N-duals}) \\ (ii)\ \forall C^\Theta \exists C^\Delta (C^\Delta \text{ and } C^\Theta \text{ are N-duals}) \end{cases}$$

$$\textit{P-duality}\begin{cases} (i)\ \forall C^\Delta \exists C^\Theta (C^\Delta \text{ and } C^\Theta \text{ are P-duals}) \\ (ii)\ \forall C^\Theta \exists C^\Delta (C^\Delta \text{ and } C^\Theta \text{ are P-duals}) \end{cases}$$

We would argue that these conjectures make precise the question whether there is really anything that can be accomplished using an alignment-based 'distance' score, which cannot by accomplised via an alignment-based 'similarity' score, and vice versa. For example, if it turns out that N-duality does not hold, then categorisation outcomes via k-NN may not reproducible by an interchange of distance and similarity, and if P-duality does not hold, hierarchical clustering outcomes may likewise not be reproducible. Note that what is at stake is whether parameter settings for alignment-based distances and similarities can always be found such as to re-express the same orderings. This is distinct from seeking an arbitrary conversion on the values of one of these measures, potentially replicating an ordering, because the output values may not be attainable via any parameter setting.

There have been comparable analyses of similarity and distance measures based on sets and vectors [1, 10, 11], motivated similarly by the question whether anything which can be accomplished with one or other such measure can be replicated by another such measure. In the case of alignment-based measures on sequences and trees, as far as we are aware, these notions seem not have been systematically considered and the following sections endeavour to fill that gap.

2 Alignment-Duality

The following lemma will be useful for considering the A-duality conjectures above:

Lemma 1. *For any C^Δ, and some choice δ such that $0 \le \delta/2$ and $\forall x \in \Sigma(\delta/2 \le C^\Delta(x,\lambda), \delta/2 \le C^\Delta(\lambda,x))$ let C^Θ be defined according to (i) below. For any C^Θ, and choice δ such that $0 \le \delta$ and $\forall x \in \Sigma \forall y \in \Sigma(\delta \ge C^\Theta(x,y))$ let C^Δ be defined according to (ii) below.*

$$(i) \begin{cases} C^\Theta(x,\lambda) = C^\Delta(x,\lambda) - \delta/2 \\ C^\Theta(\lambda,y) = C^\Delta(\lambda,y) - \delta/2 \\ C^\Theta(x,y) = \delta - C^\Delta(x,y) \end{cases} \quad (ii) \begin{cases} C^\Delta(x,\lambda) = C^\Theta(x,\lambda) + \delta/2 \\ C^\Delta(\lambda,y) = C^\Theta(\lambda,y) + \delta/2 \\ C^\Delta(x,y) = \delta - C^\Theta(x,y) \end{cases}$$

then in either case, for any $\sigma : S \mapsto T$

$$\Delta(\sigma) + \Theta(\sigma) = \delta/2 \times (\sum_{s \in S}(1) + \sum_{t \in T}(1)) \ . \tag{3}$$

Proof. If defining C^Θ from C^Δ by (i), by the choice of δ we have the nonnegativity of $C^\Theta(x,\lambda)$ and $C^\Theta(\lambda,y)$. If defining C^Δ from C^Θ by (ii), by the choice of δ, we have the nonnegativity of all entries in C^Δ. Then whether defining C^Θ from C^Δ by (i), or C^Δ from C^Θ by (ii), it is straightforward to show (see Appendix)

$$\Delta(\sigma) + \Theta(\sigma) = \delta/2 \times (2|\mathcal{M}| + |\mathcal{D}| + |\mathcal{I}|) \ . \tag{4}$$

But then (3) follows since $2|\mathcal{M}| + |\mathcal{D}| + |\mathcal{I}| = \Sigma_{s \in S}(1) + \Sigma_{t \in T}(1)$ □

Theorem 1. *A-duality (i) and (ii) hold*

Proof. Immediate given the constant summation property of (3) □

Example 1 revisited The C^Θ of Example 1 can be seen as derived from the C^Δ with $\delta = 2$, and below are shown some outcomes for a few choices of δ:

$$\begin{array}{rlll} & \Delta \text{ with } C^\Delta(x,\lambda) = 1 & C^\Delta(x,x) = 0 & C^\Delta(x,y) = 1, \text{ otherwise} \\ (\delta = 2) & \Theta \text{ with } C^\Theta(x,\lambda) = 0 & C^\Theta(x,x) = 2 & C^\Theta(x,y) = 1, \text{ otherwise} \\ (\delta = 1) & \Theta \text{ with } C^\Theta(x,\lambda) = 0.5 & C^\Theta(x,x) = 1 & C^\Theta(x,y) = 0, \text{ otherwise} \\ (\delta = 0) & \Theta \text{ with } C^\Theta(x,\lambda) = 1 & C^\Theta(x,x) = 0 & C^\Theta(x,y) = -1, \text{ otherwise} \end{array}$$

The property of alignment dualizability between distance and similarity (and vice versa) expressed above in Lemma 1 and Theorem 1 was essentially first proven for the case of sequence comparison in [13]. This shows that for alignment ordering, 'distance' and 'similarity' are interchangeable. In Sects. 3.1 and 3.2 we turn next to the other ordering duality conjectures which were noted in Sect. 1.1, recalling that the N-duality conjectures are relevant to k-NN classification, and P-duality conjectures to hierarchical clustering. Before turning to that, in the following section some further A-dualizing conversions are noted.

2.1 Additional Conversions Generating A-duals

Concerning A-duals, there are besides the conversions given in Lemma 1, others which also generate A-duals. Whilst the conversion in the lemma leads to a sum of alignment costs depending only on the two trees sizes: $\delta/2 \times (\Sigma_{s \in S}(1) + \Sigma_{t \in T}(1))$, there is a conversion from distance to similarity which gives a sum of alignment costs depending only on the cost of total deletion and insertion on S and T, and a conversion from similarity to distance depending only on the self-similarities of S and T.

Lemma 2. *For any C^{Δ}, for any k, let C^{Θ} be defined according to (iii) below.*

$$(iii) \begin{cases} C^{\Theta}(x,\lambda) = kC^{\Delta}(x,\lambda) \\ C^{\Theta}(\lambda,y) = kC^{\Delta}(\lambda,y) \\ C^{\Theta}(x,y) = (1-k)(C^{\Delta}(x,\lambda) + C^{\Delta}(\lambda,y)) - C^{\Delta}(x,y) \end{cases}$$

Then for any $\sigma : S \mapsto T$

$$\Delta(\sigma) + \Theta(\sigma) = (1-k) \times (\sum_{s \in S}(C^{\Delta}(s,\lambda)) + \sum_{t \in T}(C^{\Delta}(\lambda,t))) \tag{5}$$

Lemma 3. *For any C^{Θ}, for any k, let C^{Δ} be defined according to (iv) below.*

$$(iv) \begin{cases} C^{\Delta}(x,\lambda) = C^{\Theta}(x,\lambda) + kC^{\Theta}(x,x) \\ C^{\Delta}(\lambda,y) = C^{\Theta}(\lambda,y) + kC^{\Theta}(y,y) \\ C^{\Delta}(x,y) = k(C^{\Theta}(x,x) + C^{\Theta}(y,y)) - C^{\Theta}(x,y) \end{cases}$$

Then for any $\sigma : S \mapsto T$,

$$\Delta(\sigma) + \Theta(\sigma) = k \times (\sum_{s \in S}(C^{\Theta}(s,s)) + \sum_{t \in T}(C^{\Theta}(t,t))) \tag{6}$$

For the the proofs of (5) and (6), see the Appendix.

3 Neighbour and Pair Ordering

3.1 Distance to Similarity

Having seen that A-duals can always be created in both directions, attention shifts to N-duals and P-duals.

The case of using $\delta = 0$ in the (i) conversion of Lemma 1 from C^{Δ} to C^{Θ} gives non-positive values for all non-deletion, non-insertion entries in C^{Θ}, and is an especially trivial case of dualizing a distance setting C^{Δ}, with the effect that $\Theta(S,T) = -1 \times \Delta(S,T)$. Because of this, this distance-to-similarity conversion makes not only A-duals but also N-duals and P-duals.

Theorem 2. *N-duality (i) and P-duality (i) hold.*

Proof. Left to the reader

3.2 Similarity to Distance

The remaining ordering expressibility conjectures of Sect. 1.1 are *N-duality (ii)* and *P-duality (ii)*, concerning the similarity-to-distance direction. Of the remaining conjectures, *P-duality (ii)* is stronger than *N-duality (ii)*. *P-duality (i)* was provable using the distance-to-similarity conversion by negation. Concerning similarity-to-distance, in the (ii) conversion of Lemma 1 from C^{Θ} to C^{Δ}, you can only choose $\delta = 0$ if all $C^{\Theta}(x,y) \leq 0$, which is false for many natural settings of C^{Θ}. So there is not an analogous proof of *P-duality (ii)* and in fact *P-duality (ii)* does not hold

Theorem 3. *P-duality (ii) does not hold, that is, there are C^{Θ} such that there is no C^{Δ} such that C^{Θ} and C^{Δ} are P-duals.*

Proof. It is clearly possible for C^{Θ} to be such that there is no maximum value for $\Theta(S,T)$. For example for:

$$\Theta \text{ with } C^{\Theta}(a,a) = 1 \; C^{\Theta}(a,\lambda) = 1 \; C^{\Theta}(\lambda,a) = 1$$

it is clear we have $\Theta(a,a) = 1$, $\Theta(a^2,a^2) = 2$ and in general $\Theta(a^n,a^n) = n$. Let C^{Θ} be any table defining a similarity with no maximum. On the other hand, for each C^{Δ} there will be minimum value of $\Delta(S,T)$. Suppose some C^{Δ} is a P-dual to C^{Θ}. For any n let $[\Theta]_n$ (resp. $[\Delta]_n$) be the set of pairs with similarity (resp. distance) n. If C^{Δ} is a P-dual to C^{Θ}, there is some bijection between the set of similarity classes $\{[\Theta]_s\}$ and the set of distances classes of $\{[\Delta]_d\}$. Some similarity class $[\Theta]_{s_1}$ of Θ must correspond to the minimum distance class $[\Delta]_{d_0}$. Let $[\Theta]_{s_2}$ be a higher Θ class than $[\Theta]_{s_1}$. It must correspond to some Δ class $[\Delta]_{d_1}$ distinct from $[\Delta]_{d_0}$, and since $[\Delta]_{d_0}$ is the distance-minimum, this must be a higher distance class. Then for $(S_0,T_0) \in [\Delta]_{d_0}$, and $(S_1,T_1) \in [\Delta]_{d_1}$ you have $\Delta(S_0,T_0) < \Delta(S_1,T_1)$, but also $\Theta(S_0,T_0) < \Theta(S_1,T_1)$. So the supposed dual C^{Δ} does not reverse the pair-ordering of C^{Θ}. □

Of the order-relating conjectures of Sect. 1.1 the only remaining one is *N-duality(ii)*—that is the question whether every neighbour-ordering via some C^{Θ} can be replicated by a neighbour ordering via some C^{Δ}. We can show that this is false under one minor further assumption concerning the cost-table for distance, the assumption that deletion and insertion costs are *symmetric*.

Theorem 4. *There is C^Θ such that there is no C^Δ with $C^\Delta(x,\lambda) = C^\Delta(\lambda,x)$ such that C^Θ and C^Δ are N-duals.*

Proof. Let Σ be an alphabet containing a, and suppose that $C^\Theta(a,a) = \alpha > 0$, and $C^\Theta(a,\lambda) = C^\Theta(\lambda,a) = \beta \geq 0$, which are very mild assumptions concerning C^Θ. Let $S = a^2$, and let $\mathcal{T}$ be a set such that $\{a,a^3\} \subseteq \mathcal{T}$.

First consider the neighbour ordering of $\mathcal{T}$ by Θ, that is, a sequence of disjoint subsets of $\mathcal{T}$, where each subset $[\mathcal{T}]_s$, contains exactly those members T_i of $\mathcal{T}$ with $\Theta(S,T_i) = s$, so $[\mathcal{T}]_s = \{T_i | \Theta(S,T_i) = s\}$. The neighbour ordering, $N_\Theta(a^2)$, is the sequence of these sets ordered by their descending similarity values s.

For (a^2,a^3), the alignments with 2,1, and 0 a-matches have scores, $2\alpha - \beta$, $\alpha - 3\beta$ and -5β, hence alignments with two a-matches maximise Θ and $\Theta(a^2,a^3) = 2\alpha - \beta$ and $a^2 \in [\mathcal{T}]_{2\alpha-\beta}$. For (a^2,a), the alignments with 1 and 0 a-matches have scores $\alpha - \beta$ and -3β, respectively, hence alignments with one a-match maximise Θ and $\Theta(a^2,a) = \alpha - \beta$ and $a^1 \in [\mathcal{T}]_{\alpha-\beta}$.

Given $\alpha > 0$, we have $2\alpha - \beta > \alpha - \beta$, and so a^3 is in similarity class $[\mathcal{T}]_{2\alpha-\beta}$, which is strictly earlier in $N_\Theta(a^2)$ than the class $[\mathcal{T}]_{\alpha-\beta}$, to which a belongs.

Let C^Δ be an arbitrary cost table for distance—as ever, we make the nonnegativity assumptions (1). Further assume deletion and insertion entries are symmetric. Hence, for some $\alpha' \geq 0$, $C^\Delta(a,a) = \alpha'$, and for some $\beta' \geq 0$, $C^\Delta(a,\lambda) = C^\Delta(\lambda,a) = \beta'$. We need to show that the neighbour ordering by increasing distance, $N_\Delta(a^2)$, does not replicate $N_\Theta(a^2)$. We distinguish the cases (i) $2\beta' \geq \alpha'$, so $\beta' = \alpha'/2 + \kappa$, for some $\kappa \geq 0$, and (ii) $2\beta' < \alpha'$, so $\alpha' = 2\beta' + \epsilon$, for some $\epsilon > 0$. The table below gives the possible scores from the largest to smallest possible number of a matches

		(i)	(ii)
$\sigma : a^2 \mapsto a^3$	$\Delta(\sigma)$	$\Delta(\sigma)$ assuming $\beta' = \alpha'/2 + \kappa$	$\Delta(\sigma)$ assuming $\alpha' = 2\beta' + \epsilon$
2 a-matches	$2\alpha' + \beta'$	$2.5\alpha' + \kappa$ (eq. min $= \Delta(a^2,a^3)$)	$5\beta' + 2\epsilon$
1 a-matches	$\alpha' + 3\beta'$	$2.5\alpha' + 3\kappa$	$5\beta' + \epsilon$
0 a-matches	$5\beta'$	$2.5\alpha' + 5\kappa$	$5\beta'$ (min $= \Delta(a^2,a^3)$)

$\sigma : a^2 \mapsto a$	$\Delta(\sigma)$	$\Delta(\sigma)$ assuming $\beta' = \alpha'/2 + \kappa$	$\Delta(\sigma)$ assuming $\alpha' = 2\beta' + \epsilon$
1 a-matches	$\alpha' + \beta'$	$1.5\alpha' + \kappa$ (eq. min $= \Delta(a^2,a)$)	$3\beta' + \epsilon$
0 a-matches	$3\beta'$	$1.5\alpha' + 3\kappa$	$3\beta'$ (min $= \Delta(a^2,a)$)

In case (i), $\Delta(a^2,a^3) = 2.5\alpha' + \kappa$ and $\Delta(a^2,a) = 1.5\alpha' + \kappa$. Thus a belongs to the distance class $[\mathcal{T}]_{1.5\alpha'+\kappa}$ which is either equal to, or strictly earlier than the distance class $[\mathcal{T}]_{2.5\alpha'+\kappa}$ to which a^3 belongs (with equality for $\alpha' = 0$). In case (ii), $\Delta(a^2,a^3) = 5\beta'$ and $\Delta(a^2,a) = 3\beta'$. Thus a belongs to the distance class $[\mathcal{T}]_{3\beta'}$ which is either equal to, or strictly earlier than the distance class $[\mathcal{T}]_{5\beta'}$ to which a^3 belongs. In neither case do we replicate the descending similarity ordering, $N_\Theta(a^2)$, which had a^3 in a strictly earlier class than a. □

The assumptions made about the similarity setting C^{Θ} in the proof of the theorem were very mild, so we can say that it is practically never the case that a similarity setting C^{Θ} can be dualized by a distance setting C^{Δ} with symmetric deletions and insertion costs. If we drop the requirement that the N-dualizing C^{Δ} have $C^{\Delta}(a,\lambda) = C^{\Delta}(\lambda,a)$, then the argument does not go through. For example in case (i), with $C^{\Delta}(a,a) = 0$, the 2 a-match case for $a^2 \mapsto a^3$ scores $C^{\Delta}(\lambda,a)$, and the 1 a-match case for $a^2 \mapsto a$ scores $C^{\Delta}(a,\lambda)$. So in a distance setting which penalizes deletions *more* than insertions—so $C^{\Delta}(a,\lambda) = k + C^{\Delta}(\lambda,a)$ for some $k > 0$—it is possible to make the distance class to which a^3 belongs be $[\mathcal{T}]_{C^{\Delta}(\lambda,a)}$ and strictly earlier than the distance class to which a belongs, $[\mathcal{T}]_{k+C^{\Delta}(\lambda,a)}$. This is discussed further in Sect. 5.

4 Empirical Investigation

In [10] an investigation is undertaken of vector-based distance and similarity measures often used in information retrieval. For measures which do not produce equivalent neighbour orderings, they seek to quantify the degree to which the orderings differ, based on the Kendall-tau statistic for comparing orderings [8]. Some experiments are reported on below in which we quantify in a similar way the divergence from N-duality of some conversions which generate A-duals.

The experiments were performed on a set of 1334 (see the Appendix for further details of this data set). For a given distance setting, C^{Δ}, and A-dual similarity setting, C^{Θ}, repeatedly a tree S was chosen, and neighbour files $N_{\Delta}(S)$ and $N_{\Theta}(S)$ were computed, with $N_{\Delta}(S)$ the ordering of the remaining trees by ascending Δ, and $N_{\Theta}(S)$ the ordering by descending Θ, and then $N_{\Delta}(S)$ and $N_{\Theta}(S)$ were then compared by the *kendall-tau* measure τ (see the Appendix for the definition).

Table (a) in Fig. 2 gives a unit-cost C^{Δ} setting and then C^{Θ} settings via the (i) conversion of Lemma 1 for $2 \geq \delta \geq 0$. The plot (b) then shows for each δ the average of the τ comparison on the neighbour files, $N_{\Delta}(S)$ and $N_{\Theta}(S)$. The bottom-left corner of the plot (b) represents $\delta = 0$, the special case noted in Sect. 3.1 of the (i) conversion of Lemma 1 giving $\Theta(S,T) = -1 \times \Delta(S,T)$, and thereby an N-dual so that $\tau = 0$. Then as δ increases, there is progressively greater divergence from N-duality, until at $\delta = 2$ the τ score is 0.73, which corresponds to a tendency more towards order reversal than to replication.

Table (c) in Fig. 2 gives a C^{Θ} setting and then several C^{Δ} settings derivable by the (ii) conversion of Lemma 1 for $1 \leq \delta \leq 4$. Plot (d) then again shows for each δ the average of the τ comparison on the neighbour files, $N_{\Delta}(S)$ and $N_{\Theta}(S)$. As with plot (b) this shows again that although the similarity and distance settings are A-duals, and perfectly replicate alignment orderings, they are not N-duals and do not perfectly replicate neigbour orderings.

Theorem 3 concerned the non-replicability by distance of pair-orderings by similarity and this is illustrated in Fig. 3. The dendogram (a) shows a single-link clustering of the set of strings $\{a^5, a^4, a^3, a^2, a^1\}$, using similarity

a

		dual C^{Θ} for varying δ						
	C^{Δ}	2	1.5	1	0.5	0.2	0.1	0
(x,λ)	1	0	0.25	0.5	0.75	0.9	0.95	1
(x,x)	0	2	1.5	1	0.5	0.2	0.1	0
(x,y)	1	1	0.5	0	-0.5	-0.8	-0.9	-1

c

		dual C^{Δ} for varying δ						
	C^{Θ}	1	1.5	2	2.5	3	3.5	4
(x,λ)	0.5	1	1.25	1.5	1.75	2	2.25	2.5
(x,x)	1	0	0.5	1	1.5	2	2.5	3
(x,y)	0	1	1.5	2	2.5	3	3.5	4

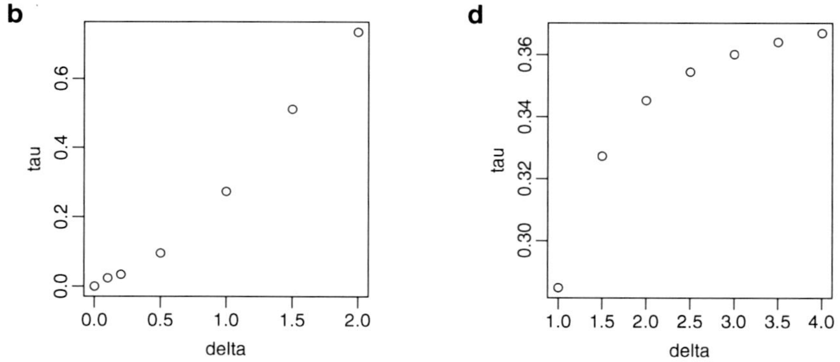

Fig. 2 (**a**) C^{Θ} settings derived from a unit-cost C^{Δ} setting by the (i) conversion of Lemma 1 as δ varies (**b**) corresponding plot of average Kendall-tau on $N_{\Delta}(S)$ and $N_{\Theta}(S)$ (**c**) C^{Δ} settings derived from a C^{Θ} setting by the (ii) conversion of Lemma 1 as δ is varied (**d**) plot of average Kendall-tau on $N_{\Delta}(S)$ and $N_{\Theta}(S)$

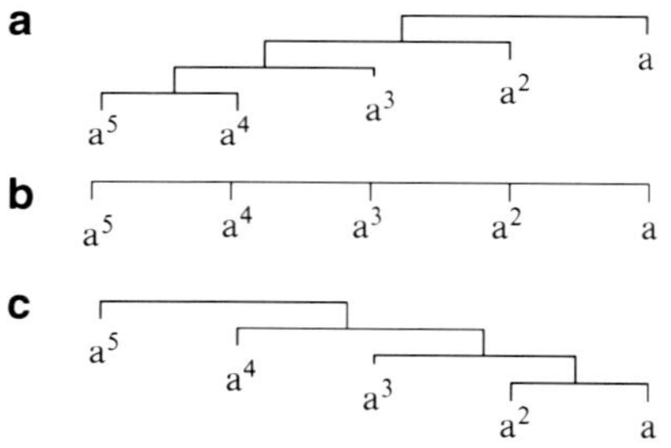

using $C^{\Theta}(a,a)=1, C^{\Theta}(a,\lambda)=1$

using $C^{\Delta}(a,a)=0, C^{\Delta}(a,\lambda)=1$

using $C^{\Delta}(a,a)=1$, $C^{\Delta}(a,\lambda)=1$
also with $0.5 \leq C^{\Delta}(a,\lambda) \leq 5.5$
(so $2C^{\Delta}(a,\lambda) \geq C^{\Delta}(a,a)$)
also with $0.1 \leq C^{\Delta}(a,\lambda) \leq 0.4$
(so $2C^{\Delta}(a,\lambda) < C^{\Delta}(a,a)$)

Fig. 3 Similarity and distance clusterings of $\{a^5,a^4,a^3,a^2,a^1\}$

with $C^{\Theta}(a,a)=1, C^{\Theta}(a,\lambda)=1$. The dendogram (b) shows the outcome with $C^{\Delta}(a,a)=0$, where all pairs (a^m,a^{m+1}) share the same minimum, $C^{\Delta}(a,\lambda)$. With $C^{\Delta}(a,a)=1$, the clustering was made for a variety of settings of the deletion/insertions, namely $0.5 \leq C^{\Delta}(a,\lambda) \leq 5.5$ and $0.1 \leq C^{\Delta}(a,\lambda) \leq 0.4$, which covers both $2C^{\Delta}(a,\lambda) \geq C^{\Delta}(a,a)$ and $2C^{\Delta}(a,\lambda) < C^{\Delta}(a,a)$. In every case the dendogram (c) was the result. Clearly the similarity-based clustering is not replicated by any of the distance-based clusterings.

5 Discussion and Comparisons

In view of the outcomes noted in Sects. 2, 3.1 and 3.2 concerning the various ordering conjectures we can say that

- Any hierarchical clustering outcome achieved via Δ can be replicated via Θ, but *not* vice versa,
- Any categorisation outcome using nearest-neighbours achieved via Δ can be replicated via Θ, but *not* vice versa,

and in this sense alignment-based 'similarity' and 'distance' comparison measures on sequences and trees are *not* interchangeable. As far as we are aware, this has not been noted before.

There are a number of papers concerning conversion between similarity to a distance measures, particularly one satisfying distance-metric axioms. In [2], Chen and Ma make a proposal concerning similarity axioms on a par with the well-known distance-metric axioms:

Distance Axioms	**Similarity Axioms** (from [2])
$D1.\Delta(S,T) = \Delta(T,S)$	$S1.\Theta(S,T) = \Theta(T,S)$
$D2.\Delta(S,T) \geq 0$	$S2.\Theta(S,S) \geq 0$
$D3.\Delta(S,V) \leq \Delta(S,T) + \Delta(T,V)$	$S3.\Theta(S,S) \geq \Theta(S,T)\Theta(S,V) + \Theta(T,T)$
$D4.\Delta(S,T) = 0$ iff $S = T$	$S4.\Theta(S,T) + \Theta(T,V) \leq$
	$S5.\Theta(S,S) = \Theta(T,T) = \Theta(S,T)$ iff $S = T$

S4 is what they propose as the similarity analog to the well-known triangle inequality D3 for distances and they define conversions from similarity to distance and in the other direction. They are not concerned directly with the P- and N-dual notions we have been discussing but rather with preservation of axiom satisfaction under their proposed conversions. The question arises as to the relation of their work to the claims we have made, one question being whether their conversions from similarity to distance, converting satisfaction of one kind of triangle inequality to another, perhaps also dualize neighbour or pair orderings. One of the conversions they propose is[6]:

$$\Delta(S,T) = (\Theta(S,S) + \Theta(T,T))/2 - \Theta(S,T) \tag{7}$$

Returning to the example considered in the proof of Theorem 4, it is straightforward to show that the similarity Θ defined there satisfies the above similarity axioms, so that a conversion according to (7) will, following the proofs in [2], satisfy the distance axioms. So, amongst other things, satisfaction of their formulation of the triangle inequality for similarity is converted to satisfaction of the standard triangle inequality. It remains the case, however, that the derived Δ is *not an N-*

[6]See Sect. 3 of [2].

dual of the similarity Θ, for we have $\Theta(a^2,a^3) = 2\alpha - \beta$, $\Theta(a^2,a^1) = \alpha - \beta$, and applying the conversion in (7), $\Delta(a^2,a^3) = (\alpha + 2\beta)/2 = \Delta(a^2,a^1)$, putting a^3 and a^1 into the same distance class in the distance-based ordering of neighbours of a^2, $N_\Delta(a^2)$, whilst they are in different classes in the similarity-based neighbour ordering, $N_\Theta(a^2)$.

In [2], Chen and Ma are not concerned specifically with *alignment*-based measures on sequences and trees, and correspondingly the conversion in (7) is defined directly on the Θ values rather than being a conversion on a cost-table C^Θ. However, it is not hard to see that a special case with $k = 1/2$ of the conversion given in Lemma 3 generates the relation between Θ and Δ of (7). Thus given a similarity setting C^Θ defining a similarity Θ which satisfies their similarity axioms, it is possible to find distance setting C^Δ which defines a distance Δ coinciding with the values generated by the conversion in (7). In [14], in work on 'global' and 'local' similarity measures on sequences (with gap functions), Spiro and Macura proposed a conversion of the parameters of an alignmment-based similarity measure to the parameters of a distance measure, inducing essentially the relation (7) between the measures. As with [2], a motivation is to demonstrate a conversion from a similarity, Θ, to a distance, Δ, satisfying the distance-metric axioms, in particular the triangle-inequality. We conjecture that the conversion of Lemma 3 is the specialisation of the Spiro and Macura conversion to the case without essential use of gap functions.

In [2, 14] a motivation was to convert a similarity to a distance satisfying the triangle inequality (D3 above). In [15], Stojmirovic and Yu share a similar motivation, and like [2] study 'global' and 'local' similarity measures on sequences (with gap functions), but they drop the symmetry condition (D1) and derive a so-called *quasi-metric*. Instantiating to the case of 'global' similarity, without gap functions, their proposal amounts to the following conversion[7]:

$$\forall x,y \in \Sigma \begin{cases} C^\Delta(x,y) = C^\Theta(x,x) - C^\Theta(x,y) \\ C^\Delta(x,\lambda) = C^\Theta(x,x) + C^\Theta(x,\lambda) \quad C^\Delta(\lambda,x) = C^\Theta(\lambda,x) \end{cases}$$

and they prove that the following relationship between Δ and Θ is induced (under particular assumptions concerning C^Θ)

$$\Delta(S,T) = \Theta(S,S) - \Theta(S,T) \tag{8}$$

When discussing Theorem 4 we noted that by having deletion costs exceed insertion costs—$C^\Delta(a,\lambda) = k + C^\Delta(\lambda,a)$ for some $k > 0$—it was possible to make the distance classes to which a^3 and a belong be $[\mathcal{T}]_{C^\Delta(\lambda,a)}$ and $[\mathcal{T}]_{k+C^\Delta(\lambda,a)}$, and thus in the same order (by ascending distances) as the similarity classes to which they belong, $[\mathcal{T}]_{2\alpha-\beta}$ and $[\mathcal{T}]_{\alpha-\beta}$ (by descending similarity). Looking at the conversion proposed by Stojmirovic and Yu, it can be seen that this is an instance of $C^\Delta(x,\lambda) = k + C^\Theta(x,\lambda)$, with $k = C^\Theta(x,x)$, given that $C^\Theta(x,\lambda) = C^\Theta(\lambda,x) = C^\Delta(\lambda,x)$. On the particular example considered in the proof of Theorem 4, we obtain

[7] See Sect. 4, Corollary 4.7 of their paper.

$\Theta(a^2,a^3) = 2\alpha - \beta$, $\Theta(a^2,a^1) = \alpha - \beta$ alongside $\Delta(a^2,a^3) = \beta$, $\Delta(a^2,a^1) = \alpha + \beta$. Not only does the conversion proposed by Stojmirovic and Yu deal with this particular example, but it is clear from the relation in (8) that this equation (unlike (7)) quite generally gives a distance which *is an N-dual* of the similarity. Thus the conversion by Stojmirovic and Yu shows how, in many cases,[8] to obtain an N-dualizing distance, it is *sufficient* to derive a distance table C^{Δ} which is asymmetric. Theorem 4 can be seen as showing that such an asymmetry is *necessary* in order to find an N-dualizing distance from a similarity.

Our findings on the various order-relating conjectures concern notions with specific, though widely used, definitions (Definitions 1–4). There are other closely related notions, and the corresponding questions concerning these have not been addressed. One variant is *stochastic*: in a stochastic similarity, probabilities are assigned to aspects of a mapping and *multiplied* [3, 12]. These are A-, N- and P-dualisable to distance. This is because, under a logarithmic mapping, these stochastic variants can be exactly simulated by a similarity as we have defined it. In the resulting table, all $C^{\Theta}(x,y) \leq 0$, allowing the (ii) conversion of Lemma 1 to define a C^{Δ} choosing $\delta = 0$. There are also *normalised* variants, which we have not considered.

Appendix

Proof of (4) from Lemma 1 (4) claims that $\Delta(\sigma) + \Theta(\sigma) = \delta/2 \times (2|\mathcal{M}| + |\mathcal{D}| + |\mathcal{I}|)$, when C^{Δ} and C^{Θ} are related by the (i) or (ii) conversions of the lemma.
If defining C^{Θ} from C^{Δ} by (i), for $\Theta(\sigma)$, we have:

$$\begin{aligned}
&\sum_{(i,j)\in\mathcal{M}} [\delta - C^{\Delta}(i,j)] - \sum_{i\in\mathcal{D}} [C^{\Delta}(i,\lambda) - \delta/2] - \sum_{j\in\mathcal{I}} [C^{\Delta}(\lambda,j) - \delta/2) \\
&= \delta(|\mathcal{M}| + \frac{|\mathcal{D}|}{2} + \frac{|\mathcal{I}|}{2}) - \sum_{(i,j)\in\mathcal{M}} [C^{\Delta}(i,j)] - \sum_{i\in\mathcal{D}} [C^{\Delta}(i,\lambda)] - \sum_{j\in\mathcal{I}} [C^{\Delta}(\lambda,j)] \\
&= \frac{\delta}{2}(2|\mathcal{M}| + |\mathcal{D}| + |\mathcal{I}|) - \Delta(\sigma)
\end{aligned}$$

If defining C^{Δ} from C^{Θ} by (ii), for $\Delta(\sigma)$ we have

$$\begin{aligned}
&\sum_{(i,j)\in\mathcal{M}} [\delta - C^{\Theta}(i,j)] + \sum_{i\in\mathcal{D}} [C^{\Theta}(i,\lambda) + \delta/2] + \sum_{j\in\mathcal{I}} [C^{\Delta}(\lambda,j) + \delta/2) \\
&= \delta(|\mathcal{M}| + \frac{|\mathcal{D}|}{2} + \frac{|\mathcal{I}|}{2}) - \sum_{(i,j)\in\mathcal{M}} [C^{\Theta}(i,j)] + \sum_{i\in\mathcal{D}} [C^{\Theta}(i,\lambda)] + \sum_{j\in\mathcal{I}} [C^{\Delta}(\lambda,j)] \\
&= \frac{\delta}{2}(2|\mathcal{M}| + |\mathcal{D}| + |\mathcal{I}|) - \Theta(\sigma)
\end{aligned}$$

Hence in either case (4) holds □

[8] The proofs in [15] do require that some some conditions on the input similarity table C^{Θ} be imposed.

Proof of (5) from Lemma 2 (5) claims $\Delta(\sigma)+\Theta(\sigma)=(1-k)\times(\sum_{s\in S}(C^{\Delta}(s,\lambda))+\sum_{t\in T}(C^{\Delta}(\lambda,t)))$ when C^{Θ} and C^{Δ} are related by conversion (iii) of the lemma. For $\Theta(\sigma)+\Delta(\sigma)$ we have

$$\sum_{(i,j)\in\mathcal{M}}[(1-k)[C^{\Delta}(i,\lambda)+C^{\Delta}(\lambda,j)]-C^{\Delta}(i,j)]-\sum_{i\in\mathcal{D}}[kC^{\Delta}(i,\lambda)]-\sum_{j\in\mathcal{I}}[kC^{\Delta}(\lambda,j)]$$

$$+\sum_{(i,j)\in\mathcal{M}}[C^{\Delta}(i,j)]+\sum_{i\in\mathcal{D}}[C^{\Delta}(i,\lambda)]+\sum_{j\in\mathcal{I}}[C^{\Delta}(\lambda,j)]$$

$$=(1-k)\sum_{(i,j)\in\mathcal{M}}[C^{\Delta}(i,\lambda)+C^{\Delta}(\lambda,j)]$$

$$+(1-k)\sum_{i\in\mathcal{D}}[C^{\Delta}(i,\lambda)]+(1-k)\sum_{j\in\mathcal{I}}[C^{\Delta}(\lambda,j)]$$

$$=(1-k)(\sum_{s\in S}(C^{\Delta}(s,\lambda))+\sum_{t\in T}(C^{\Delta}(\lambda,t)))$$

The final line holds because $|S|=|\mathcal{M}|+|\mathcal{D}|$ and $|T|=|\mathcal{M}|+|\mathcal{I}|$ □

Proof of (6) from Lemma 3 (6) claims $\Delta(\sigma)+\Theta(\sigma)=k\times(\sum_{s\in S}(C^{\Theta}(s,s))+\sum_{t\in T}(C^{\Theta}(t,t)))$ for C^{Δ} and C^{Θ} related to each other by the conversion (iv) of the lemma. For $\Delta(\sigma)+\Theta(\sigma)$ we have

$$\sum_{(i,j)\in\mathcal{M}}\left[k\left[C^{\Theta}(i,i)+C^{\Theta}(j,j)\right]-C^{\Theta}(i,j)\right]$$

$$+\sum_{i\in\mathcal{D}}\left[C^{\Theta}(i,\lambda)+kC^{\Theta}(i,i)\right]+\sum_{j\in\mathcal{I}}\left[C^{\Theta}(\lambda,j)+kC^{\Theta}(j,j)\right]$$

$$+\sum_{(i,j)\in\mathcal{M}}\left[C^{\Theta}(i,j)\right]-\sum_{i\in\mathcal{D}}\left[C^{\Theta}(i,\lambda)\right]-\sum_{j\in\mathcal{I}}\left[C^{\Theta}(\lambda,j)\right]$$

$$=\sum_{(i,j)\in\mathcal{M}}\left[k\left[C^{\Theta}(i,i)+C^{\Theta}(j,j)\right]\right]+\sum_{i\in\mathcal{D}}\left[kC^{\Theta}(i,i)\right]+\sum_{j\in\mathcal{I}}\left[kC^{\Theta}(j,j)\right]$$

$$=k(\sum_{s\in S}(C^{\Theta}(s,s))+\sum_{t\in T}(C^{\Theta}(t,t)))$$

Again the final line holds because $|S|=|\mathcal{M}|+|\mathcal{D}|$ and $|T|=|\mathcal{M}|+|\mathcal{I}|$ □

Definition of Kendall-Tau (with Ties)

Let N^1 and N^2 be two assignments of ranks to the same set of objects, U (with the possibility of ties). Where $\mathcal{P}$ is the set of all two-element sets of distinct objects from U, define a penalty function p on any $\{T_i,T_j\}\in\mathcal{P}$, such that (1) $p(\{T_i,T_j\})=1$ if the order in N^1 is the reverse of the order in N^2, (2) $p(\{T_i,T_j\})=0.5$ if there is a tie in N^1 but not in N^2 or vice versa and (3) $p(\{T_i,T_j\})=0$ otherwise. The Kendall-Tau distance (with ties) between N^1 and N^2, $\tau(N^1,N^2)$, is $\sum_{\{T_i,T_j\}\in\mathcal{P}}[p(\{T_i,T_j\})]\times\frac{2}{m\times(m-1)}$

Details of the Data Set for Kendall-Tau Experiments

Section 4 reports experiments quantifying the difference between neighbour files computed by distance and similarity, when the two are related by the conversion in Lemma 1. The trees represent syntax structures and originate in a data set which was used in a shared-task on identifying inter-node semantic dependencies [6]. See [4] for download information concerning this data. The nodes in these trees have multi-part labels. In Table (a) in Fig. 2, these are treated simply as identical or not, whereas for Table (c), the base line similarity compares node labels via $C^{\Theta}(x,y) = 1 - ham(x,y)$, where $ham(x,y)$ is the standard hamming distance; the table thus shows the extreme values of $C^{\Theta}(x,y)$ and $C^{\Delta}(x,y)$.

Acknowledgements This research is supported by the Science Foundation Ireland (Grant 07/CE/I1142) as part of the Centre for Next Generation Localisation (www.cngl.ie) at Trinity College Dublin.

References

1. Batagelj, V., Bren, M.: Comparing resemblance measures. J. Classif. **12**(1), 73–90 (1995)
2. Chen, S., Ma, B., Zhang, K.: On the similarity metric and the distance metric. Theoret. Comput. Sci. **410**(24–25), 2365–2376 (2009)
3. Emms, M.: On stochastic tree distances and their training via expectation-maximisation. In: Proceedings of ICPRAM 2012 International Conference on Pattern Recognition Application and Methods. SciTePress (2012)
4. Emms, M., Franco-Penya, H.: Data-set used in Kendall-Tau experiments. www.scss.tcd.ie/Martin.Emms/SimVsDistData September 8th (2011)
5. Gusfield, D.: Algorithms on Strings, Trees, and Sequences. Cambridge University Press, Cambridge (1997)
6. Haji, J., Ciaramita, M., Johansson, R., Kawahara, D., Meyers, A., Nivre, J., Surdeanu, M., Xue, N., Zhang, Y.: The conll-2009 shared task: Syntactic and semantic dependencies in multiple languages. In: Proceedings of the 13th Conference on Computational Natural Language Learning (CoNLL-2009). OmniPress (2009)
7. Herrbach, C., Denise, A., Dulucq, S., Touzet, H.: Alignment of rna secondary structures using a full set of operations. Technical Report 145, LRI (2006)
8. Kendall, M.G.: The treatment of ties in ranking problems. Biometrika **33**(3), 239–251 (1945)
9. Kuboyama, T.: Matching and learning in trees. PhD thesis, Graduate School of Engineering, University of Tokyo (2007)
10. Lesot, M.J., Rifqi, M.: Order-based equivalence degrees for similarity and distance measures. In: Proceedings of the Computational Intelligence for Knowledge-Based Systems Design, and 13th International Conference on Information Processing and Management of Uncertainty. IPMU'10, pp. 19–28. Springer, Berlin (2010)
11. Omhover, J.F., Rifqi, M., Detyniecki, M.: Ranking invariance based on similarity measures in document retrieval. In: Adaptive Multimedia Retrieval, pp. 55–64 Elsevier (2005)
12. Ristad, E.S., Yianilos, P.N.: Learning string edit distance. IEEE Trans. Pattern Recogn. Mach. Intell. **20**(5), 522–532 (1998)
13. Smith, T.F., Waterman, M.S.: Comparison of biosequences. Adv. Appl. Math. **2**(4), 482–489 (1981)
14. Spiro, P.A., Macura, N.: A local alignment metric for accelerating biosequence database search. J. Comput. Biol. **11**(1), 61–82 (2004)

15. Stojmirovic, A., Yu, Y.K.: Geometric aspects of biological sequence comparison. J. Comput. Biol. **16**, 579–610 (2009)
16. Tai, K.C.: The tree-to-tree correction problem. J. ACM (JACM) **26**(3), 433 (1979)
17. Wagner, R.A., Fischer, M.J.: The string-to-string correction problem. J. Assoc. Comput. Mach. **21**(1), 168–173 (1974)
18. Zhang, K., Shasha, D.: Simple fast algorithms for the editing distance between trees and related problems. SIAM J. Comput. **18**, 1245–1262 (1989)

Automatic Annotation of a Dynamic Corpus by Label Propagation

Thomas Lansdall-Welfare, Ilias Flaounas, and Nello Cristianini

Abstract We are interested in the problem of automatically annotating a large, constantly expanding corpus, in the case where potentially neither the dataset nor the class of possible labels that can be used are static, and the annotation of the data needs to be efficient. This application is motivated by real-world scenarios of news content analysis and social-web content analysis. We investigate a method based on the creation of a graph, whose vertices are the documents and the edges represent some notion of semantic similarity. In this graph, label propagation algorithms can be efficiently used to apply labels to documents based on the annotation of their neighbours. This paper presents experimental results about both the efficient creation of the graph and the propagation of the labels. We compare the effectiveness of various approaches to graph construction by building graphs of 800,000 vertices based on the Reuters corpus, showing that relation-based classification is competitive with support vector machines, which can be considered as state of the art. We also show that the combination of our relation-based approach and support vector machines leads to an improvement over the methods individually.

Keywords Graph construction • Label propagation • Large scale • Text categorisation

1 Introduction

A standard approach to annotation of documents in a large corpus is to use content-based classifiers, e.g. Support Vector Machines (SVMs), specialised in the detection of specific topics [20]. In the case where the corpus grows over time, these classifiers

---

T. Lansdall-Welfare • I. Flaounas • N. Cristianini
Intelligent Systems Laboratory, University of Bristol, Bristol, UK
e-mail: Thomas.Lansdall-Welfare@bris.ac.uk; Ilias.Flaounas@bris.ac.uk; Nello.Cristianini@bris.ac.uk

P. Latorre Carmona et al. (eds.), *Mathematical Methodologies in Pattern Recognition and Machine Learning*, Springer Proceedings in Mathematics & Statistics 30,
DOI 10.1007/978-1-4614-5076-4_2,

are applied to all new documents. In the case where new classifications need to be added to the system, new classifiers need to be trained and added to the set. We are interested in the design of highly autonomous systems for the management of corpora, and as part of this effort we have developed the news monitoring infrastructure 'News Outlets Analysis and Monitoring system' (NOAM) [14]. As part of adding more autonomy and flexibility to that infrastructure, we have been investigating various ways to propagate annotation across documents in a corpus that does not involve training content-based classifiers. We are further interested in the situation where the annotation of a corpus improves with time, that is with receiving new labelled data. We want the accuracy of existing labels to improve with more data, where old errors in classification are possibly being amended, and if entirely new labels start being used in a data stream, the system will be able to accommodate them automatically and efficiently.

In this paper we explore the use of label propagation algorithms to label graph nodes, so that the knowledge of our system about the corpus is represented both in the topology of the graph and in the labels attached to its nodes. This also involves using scalable graph creation methods. The naïve approach to graph construction, comparing all documents to all documents or building a complete kernel matrix [28], will not work in large-scale systems due to high computational complexity. The cost of label propagation is also an important factor on the time needed to process incoming documents.

We present a method to propagate labels across documents by creating a sparse graph representation of the data, and then propagating labels along the edges of the graph. Much recent research has focused on methods for propagation of labels, taking for granted that the graph topology is given in advance [8, 18]. In reality, unless working with web pages, textual corpora rarely have a predefined graph structure. Graph construction alone has a worst case cost of $\mathcal{O}(N^2)$ when using a naïve method due to the calculation of the full N $\times$ N pairwise similarity matrix. Our proposed method can be performed efficiently by using an inverted index, and in this way the overall cost of the method has a time complexity of $\mathcal{O}(N \log N)$ in the number of documents N.

We test our approach by creating a graph of 800,000 vertices using the Reuters RCV1 corpus [22], and we compare the quality of the label annotations obtained by majority voting against those obtained by using SVMs. We chose to compare the graph-based methods to SVMs because they are considered the state of the art for text categorisation [27]. We show that our approach is competitive with SVMs, and that the combination of our relation-based approach with SVMs leads to an improvement in performance over either of the methods individually. It is also important to notice that our methods can be easily distributed to multiple machines.

1.1 Related Work

There is a growing interest in the problem of propagating labels in graph structures. Previous work by Angelova and Weikum [2] extensively studied the propagation of labels in web graphs including a metric distance between labels, and assigning weights to web links based upon content similarity in the webpage documents. Many alternative label propagation algorithms have also been proposed over the years, with the survey [31] giving an overview of several different approaches cast in a regularisation framework. A common drawback of these approaches is the prohibitively high cost associated with label propagation. A number of recent works on label propagation [8, 9, 18] concentrate on extracting a tree from the graph, using a very small number of the neighbours for each node. While many graph-based methods do not address the problem of the initial graph construction, assuming a fully connected graph is given, or simply choosing to work on data that inherently has a graph structure, there is a large number of papers dedicated to calculating the nearest neighbours of a data point. One such approximate method, NN-Descent [13], shows promising results in terms of accuracy and speed for constructing k-Nearest Neighbour graphs, based upon the principle that 'a neighbour of a neighbour is also likely to be a neighbour'. The All-Pairs algorithm [5] tackles the problem of computing the pairwise similarity matrix often used as the input graph structure in an efficient and exact manner, showing speed improvements over another inverted-list-based approach, ProbeOpt-sort [26] and well-known signature-based methods such as Locality Sensitive Hashing (LSH) [15]. In this paper we take a broader overview, considering both the task of creating a graph from text documents, and then propagating labels for text categorisation simultaneously. We are interested in an approach that can be applied to textual streams, with the previously mentioned additional benefits offered by moving away from classical content-based classifiers. This paper is an extended version of the paper [21].

2 Graph Construction

Graph construction $\mathcal{X} \to \mathcal{G}$ deals with taking a corpus $\mathcal{X} = \{x_1, \ldots, x_n\}$, and creating a graph $\mathcal{G} = (V, E, W)$, where V is the set of vertices with document x_i being represented by the vertex v_i, E is the set of edges, and W is the edge weight matrix. There are several ways the construction can be adapted, namely the choice of distance metric and the method for maintaining sparsity.

The distance metric is used to determine the edge weight matrix W. The weight of an edge w_{ij} encodes the similarity between the two vertices v_i and v_j. The choice of metric used is mostly task dependent, relying on an appropriate selection being made based upon the type of data in $\mathcal{X}$. A common measure used for text, such as cosine similarity [24], may not be appropriate for other data types, such as when dealing with histogram data where the χ^2 distance is more meaningful [30].

Typically, a method for maintaining sparsity is required since it is not desirable to work with fully connected graphs for reasons of efficiency, and susceptibility to noise in the data [19]. This can be solved by working with sparse graphs, which are easier to process. Two popular methods for achieving sparsity include k-nearest neighbour (kNN) and ϵ-neighbourhood, both utilizing the local neighbourhood properties of each vertex in the graph [7, 19, 23]. Local neighbourhood methods are important for efficiency since a data point only relies on information about other points close by, with respect to the distance metric, to determine the neighbours of a vertex. This means that no global properties of the graph need to be calculated over the entire graph each time a new vertex is added, a consideration that has implications both for the scalability and, more generally, for the parallelisation.

The first step of creating the graph usually involves calculating the pairwise similarity score between all pairs of vertices in the graph using the appropriately chosen distance metric. Many studies assume that it is feasible to create a full $N \times N$ distance matrix [19] or that a graph is already given [8, 18]. This assumption can severely limit the size of data that is managable, limited by the $\mathcal{O}(N^2)$ time complexity for pairwise calculation. Construction of a full graph Laplacian kernel, as required by standard graph labelling methods [6, 17, 32] is already computationally challenging for graphs with 10,000 vertices [18]. Jebara et al. [19] introduce β-matching, an interesting method of graph sparsification where each vertex has a fixed degree β and show an improved performance over k-nearest neighbour, but at a cost to the complexity of the solution and the assumption that a fully connected graph is given.

We can overcome the issue of $\mathcal{O}(N^2)$ time complexity for computing the similarity matrix by using an alternative method, converting the corpus into an inverted index where each term has a pointer to the documents the term appears within. The advantage of this approach is that the corpus is mapped into a space based upon the number of terms, rather than the number of documents. This assumption relies on the size of the vocabulary $|t|$ being much smaller than the size of the corpus. According to Heaps' Law, the number of terms $|t|$ appearing in a corpus grows as $\mathcal{O}(N^\beta)$, where β is a constant between 0 and 1 dependent on the text [16]. Some experiments on English text have shown that in practice β is between 0.4 and 0.6 [3,4]. The inverted index can be built in $\mathcal{O}(NL_d)$ time where L_d is the average number of terms in a document, with a space complexity of $\mathcal{O}(NL_v)$ where L_v is the average number of unique terms per document [29].

Finding the neighbours of a document is also trivial because of the inverted index structure. A classical approach is to use the Term Frequency-Inverse Document Frequency (TF-IDF) weighting [24] to calculate the cosine similarity between two documents. This can be performed in $\mathcal{O}(L_d \log |t|)$ time for each document by performing L_d binary searches over the inverted index. Assuming β from Heaps' Law is the average value of 0.5, the time complexity for finding the neighbours of a document can be rewritten as $\mathcal{O}(\frac{L_d}{2} \log N)$. Therefore, there is a total time complexity $\mathcal{O}(N + \frac{NL_d}{2} \log N)$ for building the index and finding the neighbours of all vertices in the graph. This is equivalent to $\mathcal{O}(N \log N)$ under the assumption that the average document length L_d is constant.

A further advantage of this method is that the number of edges per vertex is limited a priori, since it is infeasible to return the similarity with all documents in the inverted index for every document. This allows the construction of graphs that are already sparse, rather than performing graph sparsification to obtain a sparse graph from the fully connected graph, e.g. [19].

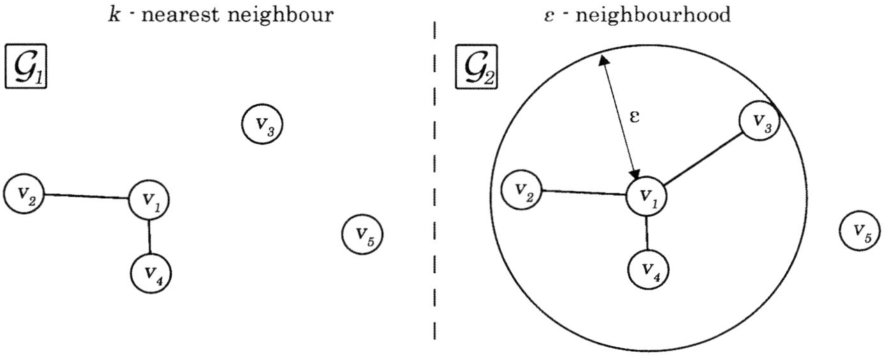

Fig. 1 Illustration of an example where two graphs, $\mathcal{G}_1$ and $\mathcal{G}_2$, are being constructed using the two methods we investigate: k-nearest neighbour and ϵ-neighbourhood. In the example, the possible edges for vertex v_1 are being considered. The k-nearest neighbour method ranks the closeness of the adjacent vertices with respect to a given similarity measure, then adds edges to the closest k vertices. For this example, $k = 2$. The ϵ-neighbourhood method adds all edges which connect v_1 to a vertex inside the large circle which visualises the radius ϵ

We investigate two popular local neighbourhood methods, k-nearest neighbour (kNN) and ϵ-neighbourhood, for keeping the graph sparse during the initial construction phase and also when new vertices are added to the graph [7, 19, 23]. Figure 1 shows intuitively how each of the methods chooses the edges to add for a given vertex. The first method, kNN, connects each vertex to the k most similar vertices in V, excluding itself. That is, for two vertices v_i and v_j, an edge is added if and only if the similarity between v_i and v_j is within the largest k results for vertex v_i. The second method we investigate, ϵ-neighbourhood, connects all vertices within a distance ϵ of each other, a similar approach to classical Parzen windows in machine learning [25]. This places a lower bound on the similarity between any two neighbouring vertices, i.e. only edges with a weight above the threshold ϵ are added to the graph. A simple way of visualising this is by drawing a sphere around each vertex with radius ϵ, where any vertex falling within the sphere is a neighbour of the vertex. While the first method fixes the degree distribution of the network, the second does not, resulting in fundamentally different topologies. We will investigate the effect of these topologies on labelling accuracy.

3 Label Propagation

Label propagation aims to use a graph $\mathcal{G} = (V, E, W)$ to propagate labels from labelled vertices to unlabelled vertices. Each vertex v_i can have multiple labels, i.e. a document can have multiple annotations, and each label is considered independently of the other labels assigned to a vertex. The labels assigned to the set of labelled vertices $\mathcal{Y}_l = \{y_1, \ldots, y_l\}$ are used to estimate the labels $\mathcal{Y}_u = \{y_{l+1}, \ldots, y_{l+u}\}$ on the unlabelled set.

Carreira-Perpinan et al. [7] suggest constructing graphs from ensembles of minimum spanning trees (MST) as part of their label propagation algorithm, with their two methods Perturbed MSTs (PMSTs) and Disjoint MSTs (DMSTs), having a complexity of approximately $\mathcal{O}(TN^2 \log N)$ and $\mathcal{O}(N^2(\log N + t))$ respectively, where N is the number of vertices, T is the number of MSTs ensembled in PMSTs, and t is the number of MSTs used in DMSTs, typically $t << \frac{N}{2}$. However, to the best of the authors' knowledge, no studies have performed experiments on constructed graphs with more than several thousand vertices, with the exception of Herbster et al. [18] who build a shortest path tree (SPT) and MST for a graph with 400,000 vertices from web pages. Herbster et al. [18] also note that constructing their MST and SPT trees using Prim and Dijkstra algorithms [11], respectively, takes $\mathcal{O}(N \log N + |E|)$ time, with the general case of a non-sparse graph having a time complexity of $\Theta(N^2)$.

In this paper we adopt Online Majority Voting (OMV) [8], a natural adaptation of the Label Propagation (LP) algorithm [32], as our algorithm for the label propagation step due to its efficiency, simplicity, and experimental performance [1]. OMV is based closely upon the locality assumption that vertices that are close to one another, with respect to a distance or measure, should have similar labels. Each vertex is sequentially labelled as the unweighted majority vote on all labels from the neighbouring vertices. The time complexity for OMV is $\Theta(|E|)$, a notable reduction from the $\mathcal{O}(kN^2)$ required for LP algorithm, where k is the neighbours per vertex. The complexity being dependent on the number of edges in the graph further benefits from the a priori limit we impose upon the maximum edges per vertex, ensuring that $|E| = bN$ for some maximum edge limit b.

4 Experiments and Evaluation

We present an experimental study of the feasibility of our approach on a large dataset, the Reuters RCV1 corpus [22]. We split the corpus into a training and test set, where the test set is the last 7 weeks of the corpus, and the training set covers everything else. The test set is further subdivided into 7 test weeks. The performance was evaluated using the F_1 Score, which is the harmonic mean of the precision and recall, a widely used performance metric for classification tasks [24]. We evaluate the performance of each method on the 7 test weeks, where all previous weeks have already been added to the graph, to simulate an online learning environment. The F_1 Scores reported are the mean performance, over the 50 most common topics, averaged over the 7 test weeks.

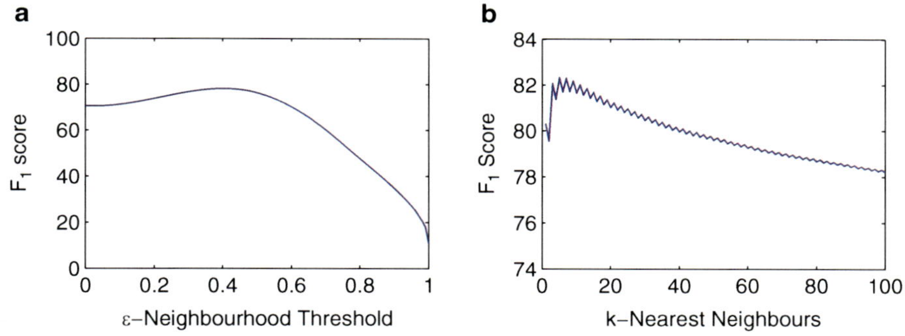

Fig. 2 F_1 performance for the 50 most common topics evaluated on the training set using LOO-CV for (a) $\epsilon = \{0, 0.01, \ldots, 1.00\}$ and (b) $k = \{1, 2, \ldots, 100\}$. It can be seen that there is a peak at $\epsilon = 0.4$ and $k = 5$

4.1 Graph-Based Content Classification

For the graph-based methods, the hyperparameters k and ϵ require careful selection in order to achieve comparable performance with current methods. This is the most expensive step as it often requires a search of the parameter space for the best value. We use Leave-One-Out Cross Validation (LOO-CV) on the training set to tune the parameters. This involves constructing graphs for a range of values of ϵ and k on the training set by iterating over all vertices, predicting the labels of the vertex based upon the majority vote of its neighbours. The predictions are checked against the true labels, with the highest performing parameter value being chosen. The performance for values of ϵ and k can be seen in Fig. 2(a) and 2(b). The best parameters for each topic individually were also recorded, allowing for a multi-parameter graph where each topic label uses a different parameter value. This could informally be thought of as each label being able to travel a certain distance along each edge. Figure 3(a) and 3(b) show the performance difference on each topic between using the general parameter values $\epsilon = 0.4$ and $k = 5$ for every topic, and using the optimal value found for each topic individually. It can be seen that for some topics a small increase in performance can be achieved, but the performance gain is minimal (with some loss for ϵ-Neighbourhood) at the expense of constructing multiple graphs, and so this approach is not considered further. Figure 4 shows a direct comparison of the graph-based methods with each other. Out of the 50 most common topics, kNN has a higher performance on 46 of the possible 50 topics. Clearly, ϵ-Neighbourhood is the weaker of the graph-based methods.

4.2 Comparison with Content-Based SVMs Performance

A comparison of the graph-based methods with the current state of the art in content-based classification was performed. The SVMs were deployed using the

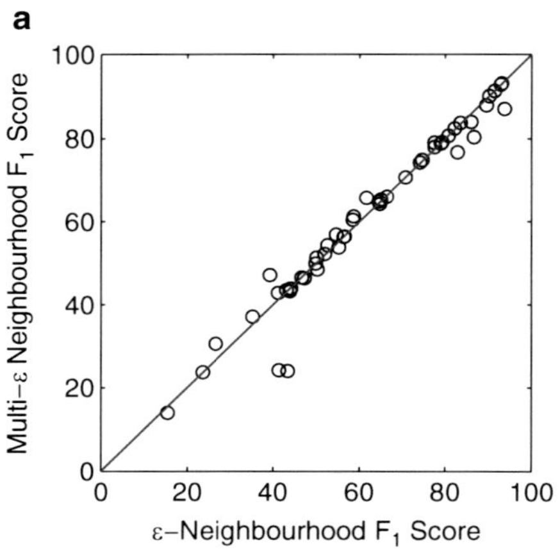

Fig. 3 Comparison of the mean F_1 Score, averaged over all test set weeks, for the graph-based methods with a single best parameter against a multi-parameter approach on the 50 most common topics. Points below the diagonal line indicate when the single parameter method achieved a higher performance, with points above the diagonal line indicating that the multi-parameter method achieved a higher performance on that topic

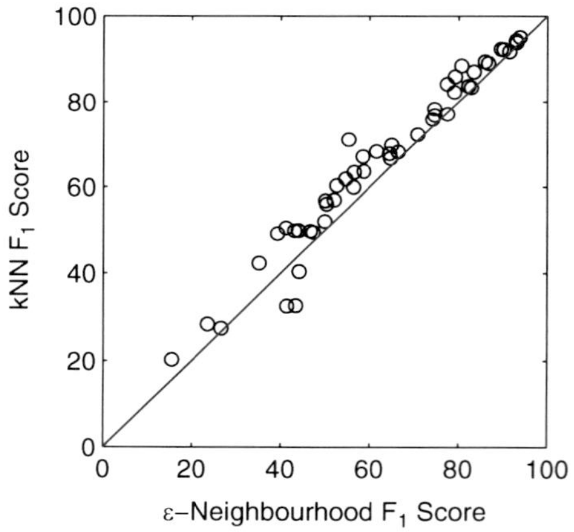

Fig. 4 Comparison of the mean F_1 Score, averaged over all test set weeks, for the graph-based methods on the 50 most common topics. Points below the diagonal line indicate when ϵ-Neighbourhood achieved a higher performance than kNN, with points above the diagonal line indicating that kNN achieved a higher performance than ϵ-Neighbourhood on that topic

LibSVM toolbox [10]. We trained one SVM per topic using the Cosine kernel, which is a normalised version of the Linear kernel [28]. For each topic, training used a randomly selected 10,000 positive examples, and a randomly selected 10,000 negative examples picked from the training set. The examples were first stemmed and stop words were removed as for the graph-based methods. The last week of the training corpus was used as a validation set to empirically tune the regularisation parameter C out of the set [0.01, 0.05, 0.1, 0.5, 1, 5, 10, 100]. For each topic, C was tuned by setting it to the value achieving the highest F_1 performance on that topic in the validation set. Figure 5(a) and 5(b) show a comparison of

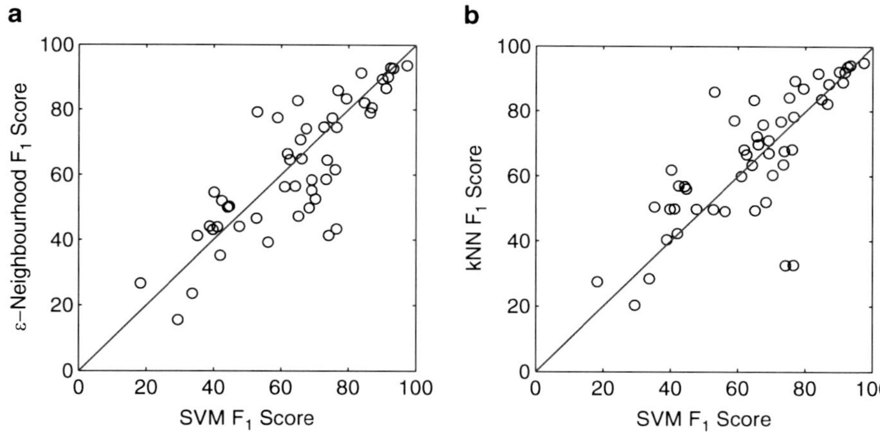

Fig. 5 Comparison of the mean F_1 Score, averaged over all test set weeks, for (**a**) ϵ-Neighbourhood and (**b**) k-NN against SVMs on the 50 most common topics. Points below the diagonal line indicate when SVMs achieved a higher performance than the graph-based method, with points above the diagonal line indicating that the graph-based method achieved a higher performance than SVMs on that topic

the graph-based methods with SVMs. Out of the 50 most common topics, SVM achieved a higher performance than ϵ-Neighbourhood on 29 topics, but only beat kNN on 19 of the topics, that is kNN performed better than SVMs on 31 out of the 50 topics. This shows that the graph-based methods are competitive with the performance of SVMs.

4.3 Building Ensembles of Graph-Based and Content-Based Approaches

Further to the comparison of the graph-based methods with SVMs, an ensemble [12] of the graph-based and content-based classification methods was evaluated. For each vertex, a majority vote for each class label c is taken by counting the supporting votes from k votes of the kNN method, supplemented with s votes from the SVMs for a total of $\upsilon = k + s$ votes. That is, each vertex has the k votes from the kNN method, but also s votes assigned by the SVMs. The number of votes from the SVM is chosen in the interval $s = [0, k+1]$. This moves the combination method from purely graph-based at $s = 0$ ($\upsilon = k$), to purely content-based at $s = k + 1$ ($\upsilon = 2k + 1$).

Given a set of p class labels $C = \{c_1, c_2, \ldots, c_p\}$, a set of n vertices $V = \{v_1, v_2, \ldots, v_n\}$, a graph matrix $A \in \{0,1\}^{n\times n}$ where $A_{i,j}$ indicates whether v_j is a neighbour of v_i, a label matrix $Y \in \{0,1\}^{n\times p}$ where $Y_{j,c}$ indicates if vertex v_j

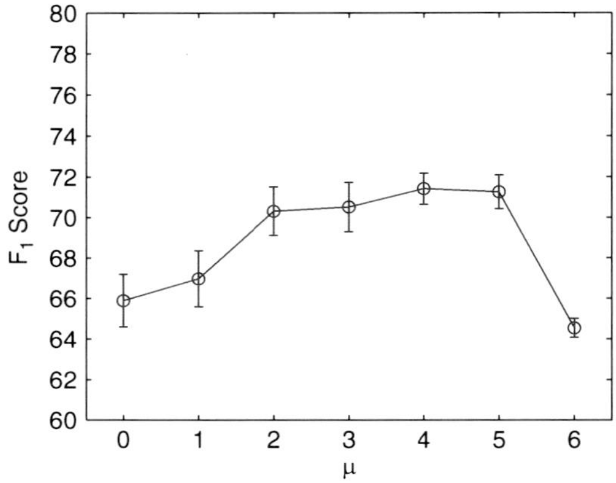

Fig. 6 Comparison of the mean F_1 Score, averaged over all test set weeks, for the combined method at different μ values on the 50 most common topics. It can be seen that the combined method offers an improvement over the *k*NN approach ($\mu = 0$) and the SVM approach ($\mu = 6$)

has class label c, an SVM assigned label matrix $S \in \{0,1\}^{n \times p}$ where $S_{i,c}$ indicates if class label c has been assigned to vertex v_i by the SVMs and a regularisation parameter $\lambda = [0,1]$, $\widetilde{Y}_{i,c}$ is the decision whether label c is to be assigned to vertex v_i. Formally, a linear combination of the methods was created as

$$\widetilde{Y}_{i,c} = \theta\left(\lambda \sum_j (A_{i,j}Y_{j,c}) + (1-\lambda)S_{i,c}\right) \tag{1}$$

$$\theta(x) = \begin{cases} 1 & \text{if } x > \frac{\upsilon}{2} \\ 0 & \text{otherwise} \end{cases} \tag{2}$$

Equation 1 can be reformulated so that it is easier to interpret by setting $\mu = \frac{1-\lambda}{\lambda}$, giving

$$\widehat{Y}_{i,c} = \theta\left(\sum_j (A_{i,j}Y_{j,c}) + \mu S_{i,c}\right) \tag{3}$$

where μ represents the number of SVM votes s in the interval $[0, k+1]$.

For our experiments, the value of μ for combining the *k*NN and SVM methods was evaluated between 0 and 6 since the *k*NN method uses $k = 5$ neighbours.

Next, we consider the best value of μ for combining the *k*NN methods and SVMs in a linear combination. Figure 6 shows the performance of the combined method averaged over the 50 most common topics for each value of μ. Out of the 50 most common topics, the combined method with $\mu = 4$ provided an improvement over the performance of both the SVM and *k*NN methods for 36 of the topics. Using $\mu = 1$ showed an improvement over both methods for the greatest number of topics, with 38 of the 50 topics seeing an improvement. The mean performance of the combined method with $\mu = 1$ is lower than for $\mu = 4$ however, indicating that when $\mu = 4$ the improvements are greater on average, even if there are slightly fewer of them.

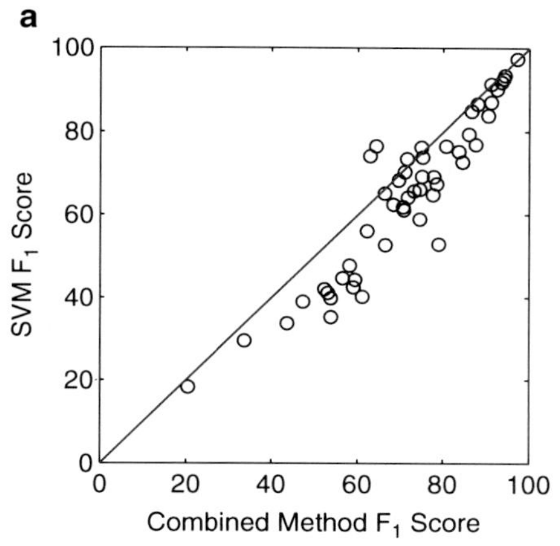

Fig. 7 Comparison of the mean F_1 Score, averaged over all test set weeks, for the combined method using $\mu = 4$ against (**a**) SVMs and (**b**) *k*NN on the 50 most common topics. Points below the diagonal line indicate when the combined method achieved a higher performance, with points above the diagonal line indicating that the individual method achieved a higher performance than the combined method on that topic

When comparing the combined method with SVM and *k*NN as seen in Fig. 7(a) and 7(b) respectively, the performance of the combined method was higher than SVM on 45 of the 50 topics and higher than *k*NN on 41 out of the 50 topics. This shows that the combined method does not only improve on SVM and *k*NN on average, but also provides an improvement for 90% and 82% of the 50 most common topics, respectively. It should be noted that in the cases where the combined method does not provide an improvement on one of the methods, it does still have a higher performance than the lowest performing method for that topic. That is, there were no cases where combining the methods gives a performance below both of the methods individually.

A summary of the overall performance of each method can be seen in Fig. 8. The ϵ-Neighbourhood method is the weaker of the two methods proposed with a performance of 62.2%, while the *k*NN method achieved a performance of 65.9%, beating the 64.5% for SVMs. Combining the *k*NN and SVM methods reached the highest performance at 71.4% with $\mu = 4$, showing that combining the relation-based and content-based approaches is an effective way to improve performance.

5 Conclusion

We have investigated a scalable method for annotating a large and growing corpus. This is achieved by efficiently creating a sparse graph and propagating the labels along its edges. In our case study the edges were created by using bag of words

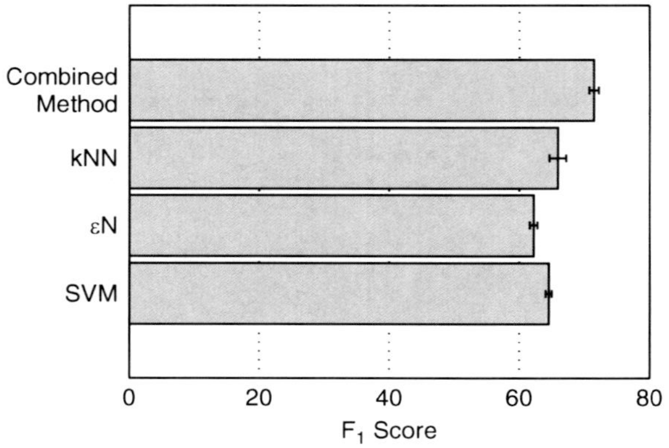

Fig. 8 Summary of the mean F_1 Score, averaged over all test set weeks, for the graph-based methods and SVMs along with the best combined method ($\mu = 4$) on the 50 most common topics. It can be seen that the graph-based methods are comparable with SVMs, with the combined method showing a further improvement. It should be noted that the performance of the combined method is slightly bias due to selecting for the best μ. ϵ-Neighbourhood has been abbreviated to ϵN

similarity, but potentially we could use any other measure that is correlated to the labels being propagated. There has been an increased theoretical interest on methods of label propagation, and some interest on how graph construction interplays with the propagation algorithms. Findings suggest that the method of graph construction cannot be studied independently of the subsequent algorithms applied to the graph [23]. We claim that label propagation has many advantages over the traditional content-based approach such as SVMs. New labels that are introduced into the system can be adopted with relative ease, and will automatically begin to be propagated through the graph. In contrast, a new SVM classifier would need to be completely trained to classify documents with the new class label. A second advantage of label propagation is that incorrectly annotated documents can be reclassified based upon new documents in a self-regulating way. That is, the graph is continuously learning from new data and improving its quality of annotation, while the SVM is fixed in its classification after the initial training period. In this paper, we have investigated two different local neighbourhood methods, ϵ-Neighbourhood and k-Nearest Neighbour, for constructing graphs for text. We have shown that sparse graphs can be constructed from large text corpora in $\mathcal{O}(N \log N)$ time, with the cost of propagating labels on the graph linear in the size of the graph, i.e. $\mathcal{O}(N)$. Our results show that the graph-based methods are competitive with content-based SVM methods. We have further shown that combining the graph-based and content-based methods leads to an improvement in performance. The proposed methods can easily be scaled out into a distributed setting using currently available open source software such as Apache Solr, or Katta, allowing a user to handle millions of texts with similarly effective performance. Research into novel ways of combining the relation and content-based methods could lead to further improvements in the categorisation performance while keeping the cost of building and propagating labels on the graph to a minimum.

Acknowledgements I. Flaounas and N. Cristianini are supported by FP7 under grant agreement no. 231495 (ComPLACS Project). N. Cristianini is supported by Royal Society Wolfson Research Merit Award. All authors are supported by Pascal2 Network of Excellence.

References

1. Ali, O., Zappella, G., De Bie, T., Cristianini, N.: An empirical comparison of label prediction algorithms on automatically inferred networks. In: Proceedings of the 1st International Conference on Pattern Recognition Applications and Methods, SciTePress, pp. 259–268 (2012)
2. Angelova, R., Weikum, G.: Graph-based text classification: learn from your neighbors. In: Proceedings of the 29th Annual International ACM SIGIR Conference on Research and Development in Information Retrieval, pp. 485–492. ACM, New York (2006)
3. Araujo, M., Navarro, G., Ziviani, N.: Large Text Searching Allowing Errors In: Proceedings of the 4th South American Workshop on String Processing (WSP'97), pp. 2-20. Carleton University Press (1997)
4. Baeza-Yates, R., Navarro, G.: Block addressing indices for approximate text retrieval. J. Am. Soc. Inform. Sci. **51**(1), 69–82 (2000)
5. Bayardo, R., Ma, Y., Srikant, R.: Scaling up all pairs similarity search. In: Proceedings of the 16th International Conference on World Wide Web, pp. 131–140. ACM, New York (2007)
6. Belkin, M., Matveeva, I., Niyogi, P.: Regularization and semi-supervised learning on large graphs. In: Learning Theory, pp. 624–638. Springer, Berlin (2004)
7. Carreira-Perpinan, M., Zemel, R.: Proximity graphs for clustering and manifold learning. In: Advances in Neural Information Processing Systems 17. NIPS-17, MIT Press (2004)
8. Cesa-Bianchi, N., Gentile, C., Vitale, F., Zappella, G.: Random spanning trees and the prediction of weighted graphs. In: Proceedings of ICML, Citeseer, OmniPress, pp. 175–182 (2010)
9. Cesa-Bianchi, N., Gentile, C., Vitale, F., Zappella, G.: Active Learning on Graphs via Spanning Trees, In NIPS Workshop on Networks Across Disciplines (2010)
10. Chang, C.C., Lin, C.J.: LIBSVM: A library for support vector machines. ACM Trans. Intell. Syst. Technol. **2**, 27:1–27:27 (2011). Software available at http://www.csie.ntu.edu.tw/~cjlin/libsvm
11. Cormen, T., Leiserson, C., Rivest, R.: Introduction to Algorithms, MIT Press and McGraw-Hill (1990)
12. Dietterich, T.: Ensemble methods in machine learning. Multiple Classifier Systems, LNCS, Vol. 1857, Springer, pp. 1–15, (2000)
13. Dong, W., Moses, C., Li, K.: Efficient k-nearest neighbor graph construction for generic similarity measures. In: Proceedings of the 20th International Conference on World Wide Web, pp. 577–586. ACM, New York (2011)
14. Flaounas, I., Ali, O., Turchi, M., Snowsill, T., Nicart, F., De Bie, T., Cristianini, N.: NOAM: news outlets analysis and monitoring system. In: Proceedings of the 2011 ACM SIGMOD International Conference on Management of Data, pp. 1275–1278. ACM, New York (2011)
15. Gionis, A., Indyk, P., Motwani, R.: Similarity search in high dimensions via hashing. In: Proceedings of the 25th International Conference on Very Large Data Bases, pp. 518–529. Morgan Kaufmann Publishers Inc., Los Altos (1999)
16. Heaps, H.: Information Retrieval: Computational and Theoretical Aspects. Academic, Orlando (1978)

17. Herbster, M., Pontil, M.: Prediction on a graph with a perceptron. Adv. Neural Inform. Process. Syst. **19**, 577 (2007)
18. Herbster, M., Pontil, M., Rojas-Galeano, S.: Fast prediction on a tree. Adv. Neural Inform. Process. Syst. **21**, 657–664 (2009)
19. Jebara, T., Wang, J., Chang, S.: Graph construction and b-matching for semi-supervised learning. In: Proceedings of the 26th Annual International Conference on Machine Learning, pp. 441–448. ACM, New York (2009)
20. Joachims, T.: Text categorization with support vector machines: Learning with many relevant features. In: Machine Learning: ECML-98, Springer, pp. 137–142 (1998)
21. Lansdall-Welfare, T., Flaounas, I., Cristianini, N.: Scalable corpus annotation by graph construction and label propagation. In: Proceedings of the 1st International Conference on Pattern Recognition Applications and Method, SciTePress, pp. 25–34 (2012)
22. Lewis, D., Yang, Y., Rose, T., Li, F.: RCV1: A new benchmark collection for text categorization research. J. Mach. Learn. Res. **5**, 361–397 (2004)
23. Maier, M., Von Luxburg, U., Hein, M.: Influence of graph construction on graph-based clustering measures. Adv. Neural Inform. Process. Syst. **22**, 1025–1032 (2009)
24. Manning, C., Raghavan, P., Schütze, H.: Introduction to Information Retrieval. Cambridge University Press, Cambridge (2008)
25. Parzen, E.: On estimation of a probability density function and mode. Ann. Math. Statist. **33**(3), 1065–1076 (1962)
26. Sarawagi, S., Kirpal, A.: Efficient set joins on similarity predicates. In: Proceedings of the 2004 ACM SIGMOD International Conference on Management of Data, pp. 743–754. ACM, New York (2004)
27. Sebastiani, F.: Machine learning in automated text categorization. ACM Comput. Surv. (CSUR) **34**(1), 1–47 (2002)
28. Shawe-Taylor, J., Cristianini, N.: Kernel Methods for Pattern Analysis. Cambridge University Press, Cambridge (2004)
29. Yang, Y., Zhang, J., Kisiel, B.: A scalability analysis of classifiers in text categorization. In: Proceedings of the 26th Annual International ACM SIGIR Conference on Research and Development in Informaion Retrieval, pp. 96–103. ACM, New York (2003)
30. Zhang, J., Marszałek, M., Lazebnik, S., Schmid, C.: Local features and kernels for classification of texture and object categories: a comprehensive study. Int. J. Comput. Vision **73**(2), 213–238 (2007)
31. Zhu, X.: Semi-supervised learning literature survey. In: Computer Science. University of Wisconsin-Madison. Madison (2007)
32. Zhu, X., Ghahramani, Z., Lafferty, J.: Semi-supervised learning using gaussian fields and harmonic functions. In: International Conference of Machine Learning, AAAI Press, vol. 20, p. 912 (2003)

Computing Voronoi Adjacencies in High Dimensional Spaces by Using Linear Programming

Juan Mendez and Javier Lorenzo

Abstract Some algorithms in Pattern Recognition and Machine Learning as neighborhood-based classification and dataset condensation can be improved with the use of Voronoi tessellation. This paper shows the weakness of some existing algorithms of tessellation to deal with high-dimensional datasets. The use of linear programming can improve the tessellation procedures by focusing on Voronoi adjacency. It will be shown that the adjacency test based on linear programming is a version of the polytope search. However, the polytope search procedure provides more information than a simple Boolean test. This paper proposes a strategy to use the additional information contained in the basis of the linear programming algorithm to obtain other tests. The theoretical results are applied to tessellate several random datasets, and also for much-used datasets in Machine Learning repositories.

Keywords Voronoi adjacencies • Nearest neighbors • Machine learning • Linear programming

1 Introduction

Pattern Recognition (PR) and Machine Learning (ML) are disciplines where the knowledge about the spatial organization of the data can improve the performance of the learning and classification procedures. Voronoi and Delaunay tessellations provide partitions of some representation spaces useful in applications concerning

J. Mendez (✉)
Departamento de Informática y Sistemas, University Las Palmas de Gran Canaria, Spain
e-mail: jmendez@dis.ulpgc.es

J. Lorenzo
Institute of Intelligent Systems, University Las Palmas de Gran Canaria, Spain
e-mail: jlorenzo@iusiani.ulpgc.es

P. Latorre Carmona et al. (eds.), *Mathematical Methodologies in Pattern Recognition and Machine Learning*, Springer Proceedings in Mathematics & Statistics 30,
DOI 10.1007/978-1-4614-5076-4_3, © Springer Science+Business Media New York 2013

the spatial organization of data collections. The tessellation process makes a partition of the space in disjunct regions or cells called Delaunay or Voronoi polytopes/polyhedra. Unfortunately Delaunay/Voronoi-based approaches have not been very successful in PR and ML (if compared with Statistical one) because the computational complexity of these methods. When the attributes that define each instance of the dataset are defined in R, every instance can be represented as a point in R^n, where n is the dimensionality of the problem. Thus, the processing of datasets with real attributes can exploit their geometrical equivalence and take advantage of many well-founded geometrical procedures.

Many PR procedures, for example Neighborhood-based Classification or Dataset Condensation, only need the adjacency relations between instances instead of full details of Voronoi or Delaunay tessellations. The Voronoi adjacency deals with the problem of checking if a pair of training instances have a common boundary, that is if both are neighbors in the Voronoi tessellation.

The Nearest Neighbor (NN) and k-NN are the most used algorithms in the family of neighborhood-based procedures. Voronoi based is only a category of search procedures in spaces that are coded by means data structures, as Delaunay/Voronoi or other spatial related threes [24]. The k parameter in k-NN is usually chosen by means of a cross-validation process over the training samples [14]. Instead of using a fix k value for the whole dataset, it will be useful to define a neighborhood that locally adapts to the data without the need for cross-validation [11, 19]. The *natural neighbors* for a test point $\mathbf{q}$ can be defined from the Voronoi tessellation of the training set as the set of training instances $\mathbf{p}_i$ whose Voronoi cell contains (or are adjacent to the cell containing) $\mathbf{q}$. This definition follows the previously introduced by Sibson [26] and Gupta et al. [19]. The natural neighbors are in a subset of instances that encloses or surrounds the test point.

Procedure of dataset editing, pruning or condensing are useful in ML applications where massive dataset are used to train practical classifiers, eg. SVM or Neural Networks. In such cases volumes of the training sets are drastically reduced with low or null loss in the information. The condensation procedures that are decision-boundary consistent [8, 14] based on Voronoi adjacency do not modify the boundary between classes. Therefore, any improvement in the computation of the Voronoi tesselation will imply a reduction in the computational cost of any procedure that can be obtained from this tesselation as the k-NN. In this approach of using spatial information provided by Dlaunay/Vornoy methods clustering method [21] is an agglomerative clustering algorithm which access density information by constructing a Voronoi diagram for the input samples.

The Voronoi tessellation procedure uses the metric distance to define the boundary planes between regions or cells. Metric distance, as well as vector norm, only can be used on spaces with a metric structure. However, many applications in ML deal with data collections without such a level of structured domains. One way to transform the experimental raw space in a metric space is to use the statistical Mahalanobis distance [28]. An equivalent approach is the use of an orthonormal linear transformation as performed in the Karhunen–Loewe (KL) transformation [28]. In this case, the Euclidean distance in the transformed space is equivalent to the Mahalanobis distance in the experimental space.

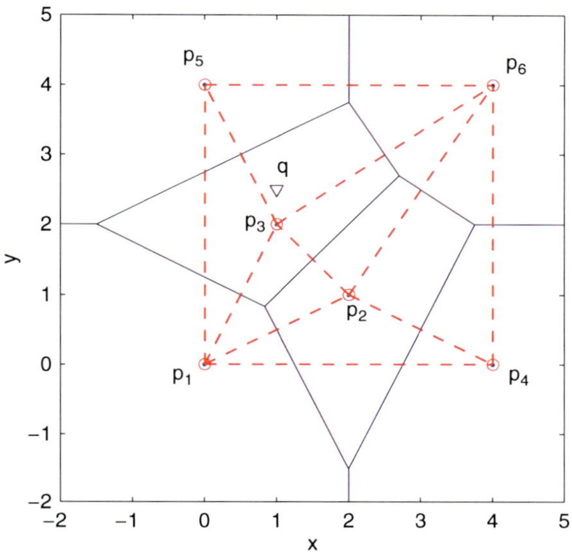

Fig. 1 A simple dataset [15], $\mathbf{P} = \{(0,0),(2,1),(1,2),\ (4,0),(0,4),(4,4)\}$, showing the Voronoi polyhedra as well as the Delaunay polytopes. The nearest neighbor of point $\mathbf{q}$ is $\mathbf{p}_3$ and its natural neighbors are: $\{\mathbf{p}_1, \mathbf{p}_2, \mathbf{p}_3, \mathbf{p}_5, \mathbf{p}_6\}$

There are several methods to compute the Voronoi and its dual the Delaunay tessellations [9, 25, 27]. Perhaps one of the more successful approaches is the one based on representation in an extended space by mapping the instances in R^{n+1}, and attempting to search their convex hull. The projection of the solution in R^n generates the tessellation. The greatest problem with the computation of Voronoi tessellation is the computational complexity. For a dataset with m instances it is in $O(m \log m)$ for 2D cases, and for a space with dimension n, it is in $O(m^{n/2})$ in the general case [6, 19], which is clearly exponential with the problem dimensionality. Figure 2 shows the results of the program *qvoronoi*, a member of the *qhull* package [6], for some UCI datasets [2]. It is highly efficient in computing low-dimensional datasets, but cannot tessellate high-dimensional datasets.

The computational complexity of Voronoi tessellations can be reduced with the use of Grabriel graphs [17], which have been used as lower cost alternatives for Voronoi adjacency [3, 4]. However, Gabriel graphs are subsets of Voronoi graphs and do not provide the full information about neighboring relations.

Computing the Voronoi or Delaunay tessellation in higher dimensional spaces can become unpractical. However, computing only the Voronoi adjacency can be done very efficiently by using Linear Programming(LP) [15]. The relationship between Voronoi and LP problems has a sound theoretical background [1, 5, 10, 16, 20] and can be continually improved with the advances in computer hardware because Linear Programming (LP) can be efficiently programmed in matrix processors as GPUs [18] and multiprocessor systems [31].

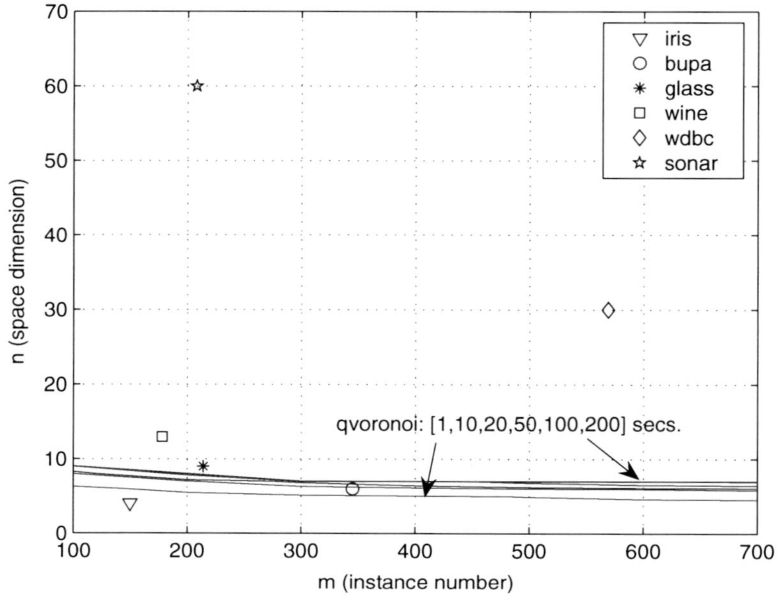

Fig. 2 Efficiency area of qvoronoi and its relation to some of the most used datasets in ML. It is very efficient, but only for low dimensionality problems

As it was stated above, any reduction in the computation of the Voronoi adjacency will imply an improvement in methods like the k-NN and condensation techniques. The aim of this paper is to present a method for an efficient computation of the Voronoi adjacency graph. This computation is based on Linear Programming and it introduces some innovations over papers previously referenced. The first one is the modification of the Voronoi adjacency test proposed by Fukuda [15] by showing that it can be reduced to the polytope search procedure. The second innovation is to show that the use of the dual problem [7, 29] of the adjacency test brings computational advantages. And last innovation, but not least, the proposal of an adjacency search strategy without backtracking. This strategy assures the computation of the correct value for all adjacency pairs without needing the computation of adjacency test for all the pairs [23].

The paper is structured as follows: firstly, the adjacency test for an instance pair is formulated, modified and transformed to its dual form. Then, the procedure of polytope search is formulated and transformed to its dual form. It will be shown that the adjacency test is a version of the polytope search. However, the polytope search procedure provides more information than a simple Boolean test. The paper proposes a strategy to use the additional information contained in the basis of the linear programming algorithm to obtain other tests. The experiments were realized with both artificial and real datasets. Real datasets with numerical features were taken from the UCI repository to allow comparisons with the results presented in this work.

2 Computation of Voronoi Adjacencies

One way to compute the Voronoi polyhedron of a dataset $\mathbf{P} = \{\mathbf{p}_1, \ldots, \mathbf{p}_m\}$ in R^n is based on the construction of an extended paraboloid representation in R^{n+1}. If $\mathbf{x} \in R^n$, the $n+1$ paraboloid coordinate is: $x_{n+1} = \sum x_1^2 + \cdots + x_n^2 = \|\mathbf{x}\|^2$. If p_{ij} are the coordinate values of $\mathbf{p}_i$, its extended representation is: $\overline{\mathbf{p}}_i = (p_{i1}, \ldots, p_{in}, \|\mathbf{p}_i\|^2)$. The set of tangent $(n+1)$-planes in every instance of the dataset generates a polyhedron whose projection in R^n is the Voronoi diagram [15]. The polyhedron is defined by the following set of linear equations:

$$2\sum_{j=1}^{n} p_{ij}x_j - x_{n+1} \leq \|\mathbf{p}_i\|^2 \qquad i = 1, \ldots, m \tag{1}$$

The adjacency of two instances $\mathbf{p}_a$ and $\mathbf{p}_b$ is verified if they have a common separating plane in R^n; therefore, the tangent planes in R^{n+1} in each instance have an intersection and a common edge. This condition is verified if a solution exits for the following linear system:

$$\begin{aligned} 2\sum_{j=1}^{n} p_{ij}x_j - x_{n+1} &\leq \|\mathbf{p}_i\|^2 \qquad i \neq a, b \\ 2\sum_{j=1}^{n} p_{aj}x_j - x_{n+1} &= \|\mathbf{p}_a\|^2 \\ 2\sum_{j=1}^{n} p_{bj}x_j - x_{n+1} &= \|\mathbf{p}_b\|^2 \end{aligned} \tag{2}$$

This feasibility test of this linear system is related to the solution of the following problem of linear programming, where $f()$ is an objective function subject to the following constraints:

$$\begin{aligned} &\text{maximize} \quad f(x_1, \ldots, x_n, x_{n+1}) \\ &2\sum_{j=1}^{n} p_{ij}x_j - x_{n+1} \leq \|\mathbf{p}_i\|^2 i \neq a, b \\ &2\sum_{j=1}^{n} p_{aj}x_j - x_{n+1} = \|\mathbf{p}_a\|^2 \\ &2\sum_{j=1}^{n} p_{bj}x_j - x_{n+1} = \|\mathbf{p}_b\|^2 \end{aligned} \tag{3}$$

This problem can be solved by introducing slack and surplus variables and using the Two-Phase Method [29]. The feasibility of this problem is obtained in the first phase by solving the next linear problem, whose goal is the minimization of the sum of all the surplus variables:

$$\text{minimize} \quad Z = s_a + s_b$$

$$2\sum_{j=1}^{n} p_{ij}x_j - x_{n+1} + s_i = \|\mathbf{p}_i\|^2 \quad i = 1,\dots,m$$

$$s_i \geq 0 \quad i = 1,\dots,m \tag{4}$$

The feasibility test for the original problem of (2) is that the optimal solution becomes null, $Z^* = 0$, equivalent to: $s_a^* = s_b^* = 0$. This problem can be modified as:

$$\text{minimize} \quad Z' = \sum_{j=1}^{n}(p_{aj} + p_{bj})x_j - x_{n+1}$$

$$2\sum_{j=1}^{n} p_{ij}x_j - x_{n+1} \leq \|\mathbf{p}_i\|^2 \quad i = 1,\dots,m \tag{5}$$

where $-Z = 2Z' - \|\mathbf{p}_a\|^2 - \|\mathbf{p}_b\|^2$, and the slack and surplus variables have been hidden. The linear programming dual of this problem is:

$$\text{minimize} \quad Z'' = \sum_{i=1}^{m} \|\mathbf{p}_i\|^2 z_i$$

$$\sum_{i=1}^{m} p_{ij}z_i = \frac{1}{2}(p_{aj} + p_{bj}) \quad j = 1,\dots,n$$

$$\sum_{i=1}^{m} z_i = 1$$

$$z_i \geq 0 \quad i = 1,\dots,m \tag{6}$$

The optimal solution of the dual must be: $Z''^* = Z'^* = \frac{1}{2}(\|\mathbf{p}_a\|^2 + \|\mathbf{p}_b\|^2)$.

2.1 Polytope Search

Voronoi polyhedra can be unbound, but a bounded polyhedron is called a polytope. Delaunay polytopes are the dual of Voronoi polyhedra. The test for Voronoi adjacency, as defined in (6), is related to the problem of the polytope search. This problem is related to find the Delaunay polytope that encloses a test point $\mathbf{q} \in R^n$: more precisely, in obtaining the subset of the dataset instances which define the polytope enclosing the test point. The polytope degree ranges from 1 to $n+1$

depending on the number of instances included, or the degree of degeneracy of the polytope. Unfortunately, not all the polytopes found are of the biggest degree of $(n+1)$; however, a lower degree provides valuable information because a k-polytope includes $k(k-1)/2$ Voronoi adjacencies. If we are trying to find the enclosing polytope of a point $\mathbf{q}$, the problem can be solved by using linear programming and finding the solution for $y_0 \in R$ and $\mathbf{y} \in R^n$ verifying[15]:

$$\text{minimize} \quad Z = y_0 + \sum_{j=1}^{n} q_j y_j$$

$$-y_0 - \sum_{j=1}^{n} p_{ij} y_j \leq \|\mathbf{p}_i\|^2 \quad i = 1, \ldots, m \tag{7}$$

The Delaunay polytope containing the test point is the one whose corresponding inequalities are satisfied as equality when the problem reaches optimal. That is, whose dual variables are not null. This linear programming algorithm has two different stop states. In the first, the enclosing polytope is found if the problem reaches the optimality. In the second one, the problem gets unbound and no solution is provided because the test point is outside the convex hull of the dataset instances. If the solution is optimal but degenerate, a k-polytope is obtained with $1 \leq k \leq n+1$. The enclosing polytope can be obtained easily by solving the dual of (7):

$$\text{minimize} \quad W = \sum_{i=1}^{m} \|\mathbf{p}_i\|^2 z_i$$

$$\sum_{i=1}^{m} \mathbf{p}_i z_i = \mathbf{q}$$

$$\sum_{i=1}^{m} z_i = 1$$

$$z_i \geq 0 \quad i = 1, \ldots, m \tag{8}$$

If the test point is outside the convex hull, the problem in (7) becomes unbound, while its dual in (8) becomes unfeasible. If the problem is not degenerate, the number of non-null problem variables $\mathbf{z} \in R^m$ is $n+1$ that define the enclosing Delaunay polytope. If the problem is degenerate, the number of non-null variables is lower. However the number of problem variables in the final basis provides some additional information: if a problem variable is null but is included in the final basis, we can infer that the k-polytope is a subset of a $(k+1)$-polytope defined by k non-null variables and this null one is also included in the basis. Therefore, knowledge of the final basis provides extra information in cases of degeneracy.

For computational purposes the polytope search procedure can be expressed as: $\mathbf{B} \leftarrow POLYTOPE(\mathbf{P}, \mathbf{q})$, where $\mathbf{B}$ is the set of instances in P included in the basis of the linear programming. The scalar $K = \text{card}(\mathbf{B})$ is an upper bound of the polytope degree, it verifies: $1 \leq k \leq K \leq n+1$. When the test point is outside the convex

Algorithm 1: Computes the Voronoi adjacency graph. The input is the dataset instances, $\mathbf{P} = \mathbf{p}_1, \ldots, \mathbf{p}_m$, and the output the graph, $\mathbf{V} = \{v_{ij}\}$.

```
procedure ADJACENCY1(P = p_1,...,p_m, V = {v_ij})
    Initialize: ∀_ij, v_ij ← 0
    for i ← 1, m−1 do
        for j ← i+1, m do                        ▷ only the upper triangular
            if v_ij = 0 then
                q ← 1/2 (p_i + p_j)              ▷ the middle point between p_i and p_j
                B ← POLYTOPE(P, q)               ▷ gets the basis of the polytope
                K ← card(B)                      ▷ by solving Equation (8)
                if K ≥ 2 then                    ▷ if degeneracy: 1 ≤ K ≤ n+1
                    for h ← 1, K−1 do
                        c ← B_h
                        for l ← h+1, K do
                            d ← B_l
                            v_cd ← 1             ▷ p_c and p_d are Voronoi neighbors
                            v_dc ← 1             ▷ and the symmetrical
```

hull, we get $K = 0$ for computational purpose. The Voronoi adjacency test for two instances in a dataset in (6) corresponds to the polytope search procedure in (8) by testing the middle point between the pair: $\mathbf{q} = \frac{1}{2}(\mathbf{p}_a + \mathbf{p}_b)$. This test point is always within the convex hull; therefore the unfeasible solution is not possible.

A Voronoi adjacency graph is constructed by taking each dataset instance as a node and the Boolean link $v_{ij} \in \{0,1\}$ as the value of the adjacency test between instances $\mathbf{p}_i$ and $\mathbf{p}_j$. The test for every middle point assures knowledge of the v_{ij} value, but in every test also other v_{hl} link values are also obtained depending on the cardinality of $\mathbf{B}$. In the best case, a number of: $n(n+1)/2+1$ links of the Voronoi adjacency graph are obtained. In the worst case only a link value is obtained: it occurs when the middle point of two instances $\mathbf{p}_i$ and $\mathbf{p}_j$ is just another instance of the dataset $\mathbf{q} = \mathbf{p}_h$. The best case happens when the middle point is within a Delaunay polytope that does not include the test pair. In this case $K = n+1$, therefore, $n(n+1)/2$ links with true values are obtained as well as a false one: $v_{ij} = 0$. The Algorithm 1 shows the proposed procedure. It initializes all the links to false values and only positive adjacencies are added throughout the following steps. Table 1 contains a trace of the computed pairs for dataset in Fig. 1, where the values for the basis variables are shown.

2.2 *Algorithm Based on Linear Programming*

The Voronoi adjacency of two points can be obtained from the polytope inclusion procedure of their middle point. We can obtain the Delaunay polytope in that a point $\mathbf{q} \in R^n$ is included based on computing the base of the following linear programming

Table 1 Computation of Voronoi neighbors for the dataset in Fig. 1

Pair	q		z_1	z_2	z_3	z_4	z_5	z_6	k	K	Links	N_s
1,2	1.0	0.5	0.50	0.50					2	2	v_{12}	2
1,3	0.5	1.0	0.50		0.50				2	2	v_{13}	2
1,4	2.0	0.0	0.50	0.00		0.50			2	3	v_{12}, v_{14}, v_{24}	3
1,5	0.0	2.0	0.50		0.00		0.50		2	3	v_{13}, v_{15}, v_{35}	3
1,6	2.0	2.0		0.40	0.40			0.20	3	3	v_{23}, v_{26}, v_{36}	3
2,5	1.0	2.5			0.75		0.19	0.06	3	3	v_{35}, v_{36}, v_{56}	4
3,4	2.5	1.0		0.75		0.19		0.06	3	3	v_{24}, v_{26}, v_{46}	4
4,5	2.0	2.0		0.40	0.40			0.20	3	3	v_{26}, v_{23}, v_{36}	3

The Algorithm 1 takes 8 tests to compute the 15 adjacent pairs, and used $N_s = 24$ iterations of Simplex Dual algorithm. The k parameter is the polytope degree, whereas K is the number of **z** variables in the basis. The filled-in **z** variables are those in the basis

problem that provides $\mathbf{Z} = \{z_1, \ldots, z_m\} \in \mathbf{1}^m$. The problem in (8) can be rewritten as: $\min \sum_{i=1}^{m} |\mathbf{p}_i - \mathbf{q}|^2 z_i = \min \sum_{i=1}^{m} |\mathbf{p}_i|^2 z_i - |\mathbf{q}|^2$:

$$
\begin{aligned}
\min \quad & \sum_{i=1}^{m} |\mathbf{p}_i|^2 z_i - |\mathbf{q}|^2 \\
st \quad & \sum_{i=1}^{m} \mathbf{p}_i z_i = \mathbf{q} \\
& \sum_{i=1}^{m} z_i = 1 \qquad z_i \geq 0
\end{aligned} \tag{9}
$$

where **Z** has a base defining the Delaunay polytope containing at most the $n+1$ points whose $z_i \in [0,1]$ values are not null. The term $|\mathbf{q}|^2$ is a constant value that can be avoided. The multidimensional dataset containing m points in an n-dimensional space is represented by the matrix **P** of dimension $n \times m$, and **Q** is a n-dimensional vector:

$$
\mathbf{P} = [\mathbf{p}_1 \cdots \mathbf{p}_m] = \begin{bmatrix} p_{11} & \cdots & p_{m1} \\ \vdots & \ddots & \vdots \\ p_{1n} & \cdots & p_{mn} \end{bmatrix} \tag{10}
$$

$$
\mathbf{q} = \begin{bmatrix} q_1 \\ \vdots \\ q_n \end{bmatrix} \tag{11}
$$

We use the Standard Dual Simplex (SDS) algorithm in this paper. According to Yarmish and Slyke [31], Revised algorithms take better advantage of sparsity in problems, while Standard algorithms are more effective for dense problems such as addressed in this paper. The LP problem of the polytope inclusion can be represented by the matrix **T**, the tableau of the LP problem. It is obtained by transforming the equality equations in (9) in pairs of inequality ones.

$$
\mathbf{T} = \left|\begin{array}{ccc|ccc|c}
|\mathbf{p}_1|^2 & \cdots & |\mathbf{p}_m|^2 & 0 & \cdots & 0 & |\mathbf{q}|^2 \\
\hline
p_{11} & \cdots & p_{m1} & & & & q_1 \\
\vdots & \ddots & \vdots & & & & \vdots \\
p_{1n} & \cdots & p_{mn} & & \mathbf{I} & & q_n \\
1 & \cdots & 1 & & & & 1 \\
\cline{1-3}\cline{7-7}
-p_{11} & \cdots & -p_{m1} & & & & -q_1 \\
\vdots & \ddots & \vdots & & & & \vdots \\
-p_{1n} & \cdots & -p_{mn} & & & & -q_n \\
-1 & \cdots & -1 & & & & -1
\end{array}\right| \qquad (12)
$$

The matrix $\mathbf{I}$ is the $2(n+1)$ order identity matrix. The matrix $\mathbf{T}$ has $2(n+1)+1$ rows and $m+2(n+1)+1$ columns. The solving of the Simplex algorithms requires three subtasks related to find the pivot, which is related to the leaving and entering variables in the base of the problem. The Simplex and Dual Simplex forms differ in the order of search for these variables. Our problem is related to the Dual forms, in that the subtasks are: find the Leaving Variable (LV) in the base, find the Entering Variable (EV) in the base, and Normalize (N) the matrices of the problem according to the pivot column and row, which can be efficiently computed in Multi-core and Multi-GPU systems [22].

2.3 *Gabriel Adjacency*

Gabriel adjacency is a subset of Voronoi adjacency, its definition resembles the general definition of Delaunay polytope. A set of $(n+1)$ instances defines a Delaunay polytope if the n-sphere that they describe has no instance into. While, two instances are Gabriel neighbors [13] if no other instance is included in the n-sphere that is centered in the middle point: $\frac{1}{2}(\mathbf{p}_a+\mathbf{p}_b)$ and has a radius: $\frac{1}{2}||\mathbf{p}_a-\mathbf{p}_b||$, that is:

$$
||\mathbf{p}_i - \frac{1}{2}(\mathbf{p}_a+\mathbf{p}_b)|| \geq \frac{1}{2}||\mathbf{p}_a-\mathbf{p}_b|| \qquad \forall i \neq a,b \qquad (13)
$$

Based on: $||\mathbf{u}-\mathbf{v}||^2 = ||\mathbf{u}||^2+||\mathbf{v}||^2-2\mathbf{u}\cdot\mathbf{v}$, it can be simplified as:

$$
\mathbf{p}_i\cdot\mathbf{p}_i - \mathbf{p}_i\cdot\mathbf{p}_a - \mathbf{p}_i\cdot\mathbf{p}_b + \mathbf{p}_a\cdot\mathbf{p}_b \geq 0 \quad \forall i \neq a,b \qquad (14)
$$

The Delaunay test, which involves $n+1$ instances to define the sphere, is more expensive than the Gabriel adjacency, which uses two instances to define a smaller sphere. If an instance pair verifies the Gabriel test, it also verifies the Voronoi neighbor test, but not the converse. This property can be used to introduce a cheaper but incomplete pre-test of adjacency.

The Gabriel test is advantageous if compared to the general Delaunay test, but this advantage is unclear when compared with Voronoi adjacency obtained with LP,

Algorithm 2: A modification of Algorithm 1 that computes all the Grabriel pre-tests previously to the polytope ones.

```
procedure ADJACENCY2(P = p_1,...,p_m, V = {v_ij})
    Initialize: ∀_ij, v_ij ← 0
    for i ← 1, m−1 do
        for j ← i+1, m do
            if GABRIEL(P, i, j) then                    ▷ tries Gabriel adjacency
                v_ij = 1
                v_ji = 1
    for i ← 1, m−1 do
        for j ← i+1, m do                               ▷ only the upper triangular
            if v_ij = 0 then
                q ← 1/2 (p_i + p_j)                     ▷ the middle point between p_i and p_j
                B ← POLYTOPE(P,q)                       ▷ gets the basis of the polytope
                K ← card(B)                             ▷ by solving Equation (8)
                if K ≥ 2 then                           ▷ if degeneracy: 1 ≤ K ≤ n+1
                    for h ← 1, K−1 do
                        c ← B_h
                        for l ← h+1, K do
                            d ← B_l
                            v_cd ← 1                    ▷ p_c and p_d are Voronoi neighbors
                            v_dc ← 1                    ▷ and the symmetrical
```

because it provides only a link value in every test. To increase the performance of the polytope-based adjacency test, one algorithm is proposed that uses the cheaper Gabriel test. The Boolean procedure GABRIEL(**P**, i, j) is used to test for the adjacency of the instances $\mathbf{p}_i$ and $\mathbf{p}_j$. In the Algorithm 2, prior to the polytope test, a pre-test is included for every instance. If the first test fails, the second one computes the pairs values.

2.4 Computing NN from Voronoi Adjacencies

The Nearest-Neighbor(NN) classification procedure of a test instance **q** in a dataset **P** is achieved by computing all the distances $d(\mathbf{q},\mathbf{p}_i)$ between the test **q** and every samples $\mathbf{p}_i \in \mathbf{P}$. If $NN()$ is such procedure, the classical NN procedure is defined as: $\mathbf{nn_q} = NN(\mathbf{P},\mathbf{q})$. The computation of NN by using Voronoi adjacencies reduces significatively that computational cost by using restricted search around a random candidate $\mathbf{n}'_\mathbf{q} \in \mathbf{P}$. The surrounding instances of this candidate are its Voronoi neighbors $V(\mathbf{n}'_\mathbf{q})$ obtained in the Learning phase of the Pattern Recognition procedure for this dataset. Algorithm 3 illustrates this process, where $V(\mathbf{a})$ includes **a** and all of its Voronoi neighbors. After the $\mathbf{n_q}$ is obtained, the *natural neighbors* of **q** are $V(\mathbf{n_q})$. These natural neighbors surround the test instance if it is into the convex hull of the dataset. Computational cost can be reduced if a list of already computed $d(\mathbf{p}_i, \mathbf{q})$ is used in the partial searches.

Algorithm 3: Computes the Nearest- Neighbor of a test instance $\mathbf{q}$ by using the Voronoi adjacency graph. The inputs are $\mathbf{V}$ the Voronoi adjacencies of $\mathbf{P}$, the test instance $\mathbf{q}$ and an initial(e.g. random) candidate $\mathbf{n}'_\mathbf{q} \in \mathbf{P}$, and the output is the Nearest-Neigbor $\mathbf{n}_\mathbf{q}$

procedure NNV($\mathbf{V}$,$\mathbf{q}$,$\mathbf{n}_\mathbf{q}$)
  Initialize: searching $\leftarrow$ 1, random $\mathbf{n}'_\mathbf{q}$
  while searching **do**
    $\mathbf{n}''_\mathbf{q} \leftarrow NN(V(\mathbf{n}'_\mathbf{q}), \mathbf{q})$
    if $\mathbf{n}''_\mathbf{q} \equiv \mathbf{n}'_\mathbf{q}$ **then**
      searching $\leftarrow$ 0 **else**
  end
  $\mathbf{n}'_\mathbf{q} \leftarrow \mathbf{n}''_\mathbf{q}$
  $\mathbf{n}_\mathbf{q} \leftarrow \mathbf{n}'_\mathbf{q}$

3 Results

A systematic test has been performed to show the strength and weakness of existing tessellation procedures. One of the more used and faster packages is the before mentioned qhull [6]. To solve several problems in computational geometry in R^n, it uses the computation of the convex hull in R^{n+1} as the kernel procedure. The family of programs based on qhull are very fast for problems with low dimensionality. However, they suffer from the curse of dimensionality when applied in high-dimensional problems such as those used in ML.

The test uses several random datasets whose instances are within the unit cube centered at the origin. The dimensions used are: $n = 2, 3, \ldots, 9, 10, 20, \ldots, 70$ and the number of instances is: $m = 100, 200, \ldots, 700$. The computation time was taken from the program *qvoronoi*, a member of the qhull package. To illustrate the obtained results for these datasets a contour plot was generated as shown in Fig. 2. Some values from 1 to 200 s are plot to illustrate the efficiency area of the procedure efficiency. The number of dimensions and instances of some of the most used dataset in the UCI Machine Learning Repository [2] such as *iris*, *bupa*, *glass*, *wine*, *wdbc* and *sonar* are also plotted. The figure shows that while *bupa* dataset (345 instances and dimension 6) can be effectively tessellated in 5.76 s. on a test computer (Intel Pentium M, 1.6 Ghz and 1 GB of RAM), the *glass* dataset (214 instances and dimension 9) took several hours to complete. A practical conclusion was obtained, that is datasets with a dimension greater than eight *cannot* be tessellated effectively with this procedure.

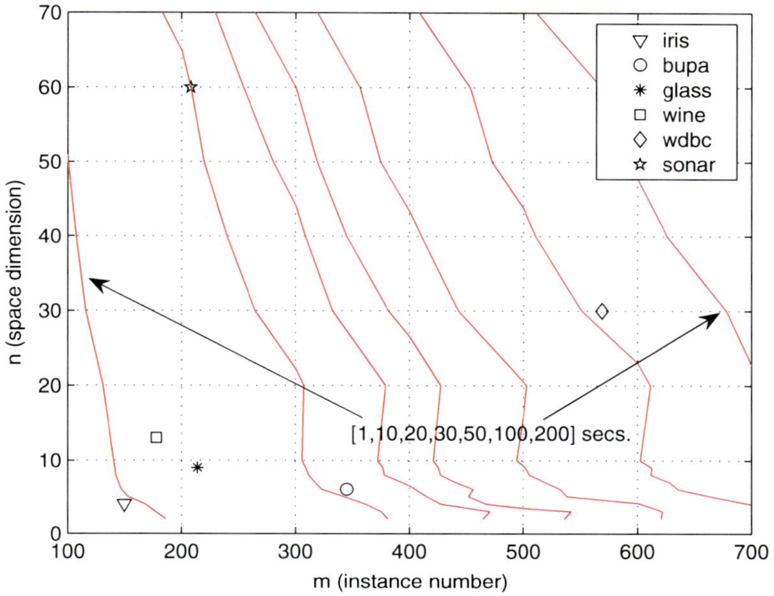

Fig. 3 Efficiency area of the tessellation procedure based on polytope search. The covered area is more extensive that the covered by qhull and includes the datasets. However, for low-dimensional problems qhull significatively outperforms it

The Algorithm 1 defines how to compute the Voronoi adjacencies, which are coded as a graph. The Boolean links $v_{ij} \in \{0,1\}$ are symmetric: $v_{ij} = v_{ji}$, therefore, only the upper triangular is computed. The Algorithm was implemented in C++ using double precision real numbers. The same systematic test that had been conducted for the qvoronoi was performed for the implementation of the Algorithm 1. The contour plot of the efficiency is shown in Fig. 3. The entire range of the UCI datasets is covered in the range of 200 s in this test computer. The efficiency area seems to cover a more extended area in the n vs. m plane, which allows to cover a wide range of practical ML applications. In low-dimensional datasets, qhull significatively outperforms the proposed implementation, but for $n \geq 8$, it is outperformed.

A performance factor is defined about how many middle point tests are required to obtain all the adjacency links of a dataset. The factor is defined as:

$$\gamma = \frac{2N_{\text{test}}}{m(m-1)} \tag{15}$$

where N_{test} is the number of tested pairs necessary to achieved the computation of all Voronoi adjacencies. It depends on each dataset, and in general would have a general dependence on m and n. Low factor values are equivalent to high tessellation performance, because its inverse provides the average number of adjacency relations obtained for each test. The tessellation cost not only depends on the size of dataset m and n, but it also depends on the distribution of instances. Table 2 contains the computation time for the ML Repository datasets, as well as the N_{test} and the Simplex iterations used. Normalized coordinates are used after the KL transformation because raw data coordinates are meaningless when used in a metric distance.

It should be mentioned that the plotted points of each of the ML Repository dataset in Fig. 3 are only qualitative because the plotted background data are related to random datasets. No qvoronoi values are available for glass dataset and larger, because these tests had not finished after several hours of computation. Therefore, they are not comparable for practical purposes. These data are computed on an Intel Xeon, 3.06 Ghz, 512K in L2 cache, and 1.5GB of RAM. The last column contains the T_2 computed for the Algorithm 2, only slight differences are detected between the two Algorithms.

The cost analysis of proposed procedure depends on the analysis cost of the LP problem for finding a Polytope enclosing a point. The cost to obtain all the Voronoi adjacencies $C_{\text{AllAd}}(n,m)$ is:

$$C_{\text{AllAd}}(n,m) = \gamma(n,m)\frac{m(m-1)}{2}C_{\text{Poly}}(n,m) \tag{16}$$

where $\gamma(n,m)$ is the fraction of the $m(m-1)/2$ pairs that that are tested. It runs, in the considered cases of the dataset in UCI, from 0.837 for iris dataset to 0.112 for wdbc dataset. In general it would have a general dependence on m and n that future works could clarify. We think that it depends for every specific dataset, and that a general dependence as $\gamma(n,m)$ is only valid as an average for random datasets.

The cost to obtain a polytope, $C_{\text{Poly}}(n,m)$, is the cost to solve an LP problem. Although we have used in practice the Simplex Dual Algorithm for practical proposes, any of the available LP Algorithms can be used. This algorithm choice is no central of our proposal; such as future works will test the relative efficiency of other choices (Simplex based variants as well as interior methods). A founded opinion [12] is that the efficiency of good implementations of simplex-based methods and interior point methods are similar for practical applications of linear programming. However, for specific types of LP problems, it may be that one type of algorithm is better than another, but it cannot be decided without an exhaustive test.

It is very difficult to define the theoretical cost of an LP Algorithm because we have to decide between the cost for *worst-case* and the cost for *average-case* in the defined application. Although the worst-case complexity of the Simplex Algorithm is exponential in the problem dimension, it is widely known that in practice it is probably a polynomial-time [30], that is in practice the Simplex method almost always converges on real-world problems in a number of iterations that is polynomial in the problem dimension.

Table 2 Tessellation results for several datasets coded in normalized coordinates after the KL transformation: T_q the computational time used by qvoronoi, T_1 the used by the proposed Algorithm 1, N_{test} the number of middle points tested, N_s the number of Simplex iterations used to tessellate the whole dataset, and γ the performance factor

Dataset	n	m	T_q (s)	T_1 (s)	N_{test}	N_s	γ	$T_2(s)$
iris	4	150	0.04	0.503	9,349	96,844	0.837	0.515
bupa	6	345	6.71	10.136	45,891	783,100	0.773	10.162
glass	9	214	n/a	2.708	9,262	256,613	0.406	2.733
wine	13	178	n/a	0.863	3,082	69,167	0.196	0.898
wdbc	30	569	n/a	58.886	18,109	664,385	0.112	61.358
sonar	60	208	n/a	6.003	5,237	46,555	0.243	6.752

The T_2 column contains the computational time for the Algoritm 2

If the average cost of LP problem for polytope finding is on the class of $O(f(m)g(n)$, where $f(m)$ and $g(n)$ are polynomial, so we can conclude that the practical cost $C_{AllAdj}(n,m)$ falls in the class $O(m^2 f(m)g(n))$ also polynomial. That is very advantageous to qhull based approaches (which are in exponential $O(m^{n/2})$ class) for large values of the space dimensionality n, but unadvantageous for small ones.

4 Conclusions

Machine Learning applications impose unattainable goals on traditional tessellation techniques, while linear programming provides alternative approaches to perform the tessellation of high-dimensional datasets. Linear programming provides a sound theoretical background for the tessellation problem as well as an inspirational source for efficient implementations. A modification of the Voronoi adjacency test had shown that it is basically the polytope search procedure, enabling the implementation of a more efficient algorithm for high-dimensional datasets. It is more efficient than a *single* adjacency test because in each trial it provides a polytope, that is *many* adjacency values. These perform best if focusing on theγparameter, which is related to the fraction of all the all-to-all needed test. The reason for this is that, the higher dimensionality the greater the number of instances included in each polytope. This is the counterpart of the curse of dimensionality. Perhaps this would be the reason why it permits a relative good performance at high dimensionality. The qhull-based and the linear programming-based implementations are complementary because each is good in different domains. A suitable use of both algorithms can efficiently tessellate many massive datasets in Machine Learning. The use of a pre-test based on the Gabriel adjacency, which provides a faster but incomplete graph of neighboring relations, does not significatively increase performance because, while it is fast, it provides only one link value while the polytope provides several link values in each test.

References

1. Agrell, E.: A Method for examining vector quantizer structures. In: Proceeding of IEEE International Symposium on Information Theory, pp. 394 (1993)
2. Frank, A. Asuncion, A. UCI Machine Learning Repository [http://archive.ics.uci.edu/ml]. Irvine, CA: University of California, School of Information and Computer Science (2010)
3. Aupetit, M.: High-dimensional labeled data analysis with gabriel graphs. In: European Symposium on Artificial Neuron Networks, pp. 21–26 (April 2003)
4. Aupetit, M., Catz, T.: High-dimensional labeled data analysis with topology representating graphs. Neurocomputing **63**, 139–169 (2005)
5. Avis, D., Fukuda, K.: A pivoting algorithm for convex hulls and vertex enumeration of arrangements and polyhedra. Discrete Comput. Geom. **8**(3), 295–313 (1992)
6. Barber, C.B., Dobkin, D.P., Huhdanpaa, H.: The quickhull algorithm for convex hulls. ACM Trans. Math. Software **22**(4), 469–483 (1996)
7. Bazaraa, M.S., Jarvis, J.J., Sherali, H.S.: Linnear Programming and Networks Flows. Wiley, New York (1990)
8. Bhattacharya, B., Poulsen, R., Toussaint, G.: Application of proximity graphs to editing nearest neighbor decision rules. Technical Report SOCS 92.19, School of Computer Science, McGill University (1992)
9. Bowyer, A.: Computing Dirichlet tessellations. Comput. J. **24**(2), 162–166 (1981)
10. Bremner, D., Fukuda, K., Marzetta, A.: Primal-dual methods for vertex and facet enumeration. In: SCG '97: Proceedings of the Thirteenth Annual Symposium on Computational Geometry, pp. 49–56. ACM, New York (1997)
11. Chin, E., Garcia, E.K., Gupta, M.R.: Color management of printers by regression over enclosing neighborhoods. In: IEEE International Conference on Image Processing. ICIP, **2**, 161–164, (Oct 2007)
12. Dantzig, G.B., Thapa, M.N.: Linear Programming 2: Theory and Extensions. Springer, Berlin (2003)
13. Devroye, L., Gyorfi, L., Lugosi, G.: A Probabilistic Theory of Pattern Recognition, Springer Verlag (1996)
14. Duda, R., Hart, P., Stork, D.: Pattern Classification. Wiley, New York (2001)
15. Fukuda, K.: Frecuently asked questions in polyhedral computation. Technical report, Swiss Federal Institute of Technology, Lausanne, Switzerland (June 2004)
16. Fukuda, K., Liebling, T.M., Margot, F.: Analysis of backtrak algoritms for listing all vertices and all faces of convex polyhedron. Comput. Geom. **8**, 1–12 (1997)
17. Gabriel, K.R., Sokal, R.R.: A new stattistical approach to geographic variation analysis. Systemat. Zool. **18**, 259–270 (1969)
18. Greeff, G.: The revised simplex algorithm on a GPU. Technical report, Department of Computer Science, University of Stellenbosch (February 2005)
19. Gupta, M.R., Garcia, E.K., Chin, E.: Adaptive local linear regression with application to printer color management. In: IEEE Transactions on Image Process (2008)
20. Kalai, G.: Linear programming, the simplex algorithm and simple polytopes. Math. Program. **79**, 217–233 (1997)
21. Koivistoinen, H., Ruuska, M., Elomaa, T.: A voronoi diagram approach to autonomous clustering. Lecture Notes in Computer Science (4265), pp. 149–160 (2006)
22. Mendez, J.: Cooperating Multi-core and Multi-GPU in the computation of the multidimensional voronoi adjacency in machine learning datasets. In: PDPTA, pp. 717–723 (2010)
23. Mendez, J., Lorenzo, J.: Efficient computation of voronoi neighbors based on polytope Search in pattern recognition. In: Proceedings of the 1st International Conference on Pattern Recognition Applications and Methods, pp. 357–364. SciTePress (2012)
24. Navarro, G.: Searching in metric spaces by spatial approximation. VLDB J. **11**, 28–46 (2002)

25. Ramasubramanian, V., Paliwal, K.: Voronoi projection-based fast nearest-neighbor search algorithms: Box-search and mapping table-based search techniques. Digital Signal Process. **7**, 260–277 (1997)
26. Sibson, R.: A brief description of natural neighbour interpolation. In: Interpreting Multivariate Data, pp. 21–36. Wiley, New York (1981)
27. Watson, D.F.: Computing the n-dimensional tessellation with application to voronoi polytopes. Comput. J. **24**(2), 167–172 (1981)
28. Web, A.: Statistical Pattern Recognition, 2nd edn. Wiley, New York (2002)
29. Winston, W.L.: Operations Research Applications and Algorithms. Wadsworth, Belmont (1994)
30. Wright, M.H.: The interior-point revolution in optimization: History, recent developments, and lasting consequences. Bull. AMS **42**(1), 39–56 (2004)
31. Yarmish, G., van Slyke, R.: RetroLP, an implementation of the standard Simplex method. Technical report, Department of Computer and Information Science, Brooklyn College (2001)

Phase-Locked Matrix Factorization with Estimation of the Common Oscillation

Miguel Almeida, Ricardo Vigário, and José Bioucas-Dias

Abstract Phase-Locked Matrix Factorization (PLMF) is an algorithm to perform separation of synchronous sources. Such a problem cannot be addressed by orthodox methods such as Independent Component Analysis, because synchronous sources are highly mutually dependent. PLMF separates available data into the mixing matrix and the sources; the sources are then decomposed into amplitude and phase components. Previously, PLMF was applicable only if the oscillatory component, common to all synchronized sources, was known, which is clearly a restrictive assumption. The main goal of this paper is to present a version of PLMF where this assumption is no longer needed—the oscillatory component can be estimated alongside all the other variables, thus making PLMF much more applicable to real-world data. Furthermore, the optimization procedures in the original PLMF are improved. Results on simulated data illustrate that this new approach successfully estimates the oscillatory component, together with the remaining variables, showing that the general problem of separation of synchronous sources can now be tackled.

Keywords Matrix factorization • Phase synchrony • Phase-locking • Independent component analysis • Blind source separation • Convex optimization

---

M. Almeida (✉)
Institute of Telecommunications, Instituto Superior Técnico, Lisbon, Portugal

Department of Information and Computer Science, Aalto University, Finland
e-mail: malmeida@lx.it.pt

R. Vigário
Department of Information and Computer Science, Aalto University, Finland
e-mail: ricardo.vigario@aalto.fi

J. Bioucas-Dias
Institute of Telecommunications, Instituto Superior Técnico, Lisbon, Portugal
e-mail: bioucas@lx.it.pt

P. Latorre Carmona et al. (eds.), *Mathematical Methodologies in Pattern Recognition and Machine Learning*, Springer Proceedings in Mathematics & Statistics 30,
DOI 10.1007/978-1-4614-5076-4_4,

1 Introduction

Synchrony is an increasingly studied topic in modern science. On the one hand, there is an elegant yet deep mathematical framework which is applicable to many domains where synchrony is present, including laser interferometry, the gravitational pull of stellar objects, and the human brain [11].

It is believed that synchrony plays an important role in the way different sections of human brain interact. For example, when humans perform a motor task, several brain regions oscillate coherently with the muscle's electromyogram (EMG) [10, 12]. Also, processes such as memorization and learning have been associated with synchrony; several pathologies such as autism, Alzheimer's and Parkinson's are associated with a disruption in the synchronization profile of the brain; and epilepsy is associated with an anomalous increase in synchrony [14].

To infer knowledge on the synchrony of the networks present in the brain or in other real-world systems, one must have access to the dynamics of the individual oscillators (which we will call "sources"). Usually, in the brain electroencephalogram (EEG) and magnetoencephalogram (MEG), and other real-world situations, individual oscillator signals are not directly measurable; one has only access to a superposition of the sources.[1] In fact, EEG and MEG signals measured in one sensor contain components coming from several brain regions [9]. In this case, spurious synchrony occurs, as has been shown both empirically and theoretically in previous works [2]. We briefly review this evidence in Sect. 2.3.

Undoing this superposition is usually called a blind source separation (BSS) problem. Typically, one assumes that the mixing is linear and instantaneous, which is a valid and common approximation in brain signals [15] and other applications. In this case, if the vector of sources is denoted by $\mathbf{s}(t)$ and the vector of measurements by $\mathbf{x}(t)$, they are related through $\mathbf{x}(t) = \mathbf{M}\mathbf{s}(t)$ where $\mathbf{M}$ is a real matrix called the mixing matrix. Even with this assumption, the BSS problem is ill-posed: there are infinitely many solutions. Thus, one must also make some assumptions on the sources, such as statistical independence in Independent Component Analysis (ICA) [7]. However, in the case discussed in this paper, independence of the sources is not a valid assumption, because synchronous sources are highly dependent. In this paper we address the problem of how to separate these dependent sources, a problem we name Separation of Synchronous Sources, or Synchronous Source Separation (SSS). Although many possible formal models for synchrony exist (see, e.g., [11] and references therein), in this paper we use a simple yet popular measure of synchrony: the Phase Locking Factor (PLF), or Phase Locking Value (PLV).

[1] In EEG and MEG, the sources are not individual neurons, whose oscillations are too weak to be detected from outside the scalp even with no superposition. In this case, the sources are populations of closely located neurons oscillating together.

The PLF between two signals is 1 if they are perfectly synchronized. Thus, in this paper we tackle the problem of source separation where all pairs of sources have a PLF of 1.

A more general problem has also been addressed, where the sources are organized in subspaces, with sources in the same subspace having strong synchrony and sources in different subspaces having weak synchrony. This general problem was tackled with a two-stage algorithm called Independent Phase Analysis (IPA) which performed well in the noiseless case [1] and with moderate levels of added Gaussian white noise [2]. In short, IPA uses TDSEP [16] to separate the subspaces from one another. Then, each subspace is a separate SSS problem; IPA uses an optimization procedure to complete the intra-subspace separation. Although IPA performs well for the noiseless case, and for various types of sources and subspace structures, and can even tolerate moderate amounts of noise, its performance for higher noise levels is unsatisfactory. Also, in its current form, IPA is limited to square mixing matrices, i.e., to a number of measurements equal to the number of sources. It may as well return singular solutions, where two or more estimated sources are (almost) identical. On the other hand, IPA can deal with subspaces of phase-locked sources and with sources that are not perfectly phase locked [2].

In this paper we address an alternative technique, named Phase-Locked Matrix Factorization (PLMF). PLMF was originally introduced in [3], using a very restrictive assumption, of prior knowledge of the oscillation common to all the sources. The goal of this paper is to remove this restrictive assumption and to improve the optimization of the problem.

Unlike IPA, PLMF can deal with higher amounts of noise and with non-square mixing matrices (more measurements than sources). Furthermore, it only uses variables directly related to the data model, and is immune to singular solutions. PLMF is inspired on the well-known Non-negative Matrix Factorization (NMF) approach [8], which is not applicable directly to the SSS problem, because some factors in the factorization are not positive, as will be made clear below. For simplicity, we will restrict ourselves to the case where the sources are perfectly synchronized.

One should not consider PLMF as a replacement for IPA, but rather as a different approach to a similar problem: PLMF is a model-driven algorithm, whereas IPA is data-driven. As we will show, PLMF has advantages and disadvantages relative to IPA.

This paper is organized as follows. In Sect. 2 we introduce the Phase-Locking Factor (PLF) quantity which measures the degree of synchronization of two signals, and show that full synchronization between two signals has a very simple mathematical characterization. Section 3 describes the PLMF algorithm in detail. In Sect. 4 we explain how the simulated data was generated and show the results obtained by PLMF. Directions for future work are discussed in Sect. 5. Conclusions are drawn in Sect. 6.

2 Phase Synchrony

2.1 Phase of a Real-Valued Signal

In this paper we tackle the problem of Separation of Synchronous Sources (SSS). The sources are assumed to be synchronous, or phase-locked: thus, one must be able to extract the phase of a given signal. In many real-world applications, such as brain EEG or MEG, the set of measurements available is real-valued. In those cases, to obtain the phase of such measurements, it is usually convenient to construct a set of complex-valued data from them. Two approaches have been used in the literature: complex wavelet transforms [13] and the Hilbert transform [6].

In this paper we present only results on simulated data, which is directly generated as complex-valued, thus curcumventing this issue.

2.2 Phase-Locking Factor

Let $\phi_j(t)$ and $\phi_k(t)$, for $t = 1,\ldots,T$, be the time-dependent phases of signals j and k. The real-valued[2] Phase-Locking Factor (PLF) between those two signals is defined as

$$\rho_{jk} \equiv \left| \frac{1}{T} \sum_{t=1}^{T} \mathrm{e}^{\mathrm{i}[\phi_j(t)-\phi_k(t)]} \right| = \left| \left\langle \mathrm{e}^{\mathrm{i}(\phi_j-\phi_k)} \right\rangle \right|, \tag{1}$$

where $\langle\cdot\rangle$ is the time average operator, and $\mathrm{i} = \sqrt{-1}$. Note that $0 \leq \rho_{jk} \leq 1$. The value $\rho_{jk} = 1$ corresponds to two signals that are fully synchronized: their phase lag, defined as $\phi_j(t) - \phi_k(t)$, is constant. The value $\rho_{jk} = 0$ is attained if the two phases are not correlated, as long as the observation period T is sufficiently long. Values between 0 and 1 represent partial synchrony. Typically, the PLF values are stored in a PLF matrix $\mathbf{Q}$ such that $\mathbf{Q}(j,k) \equiv \rho_{jk}$. Note that a signal's PLF with itself is trivially equal to 1: thus, for all j, $\rho_{jj} = 1$.

2.3 Effect of Mixing on the PLF

The effect of a linear mixing operation on a set of sources which have all pairwise PLFs equal to 1 is now discussed. This effect has a simple mathematical characterization: if $\mathbf{s}(t)$ is a set of such sources, and we define $\mathbf{x}(t) \equiv \mathbf{M}\mathbf{s}(t)$, with $\det(\mathbf{M}) \neq 0$, then the only possibility for the observations $\mathbf{x}$ to have all pairwise PLFs

[2]"Real-valued" is used here to distinguish from other papers, where the absolute value operator is dropped, hence making the PLF a complex quantity [2].

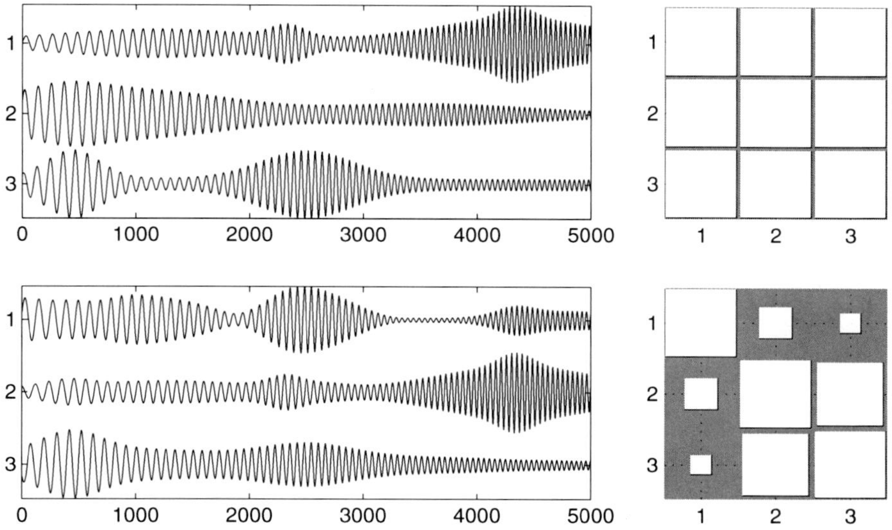

Fig. 1 *(Top row)* Three sources *(left)* and PLFs between them *(right)*. *(Bottom row)* Three mixed signals *(left)* and PLFs between them *(right)*. On the right column, the area of the square in position (i, j) is proportional to the PLF between the signals i and j. Therefore, large squares represent PLFs close to 1, while small squares represent values close to zero

equal to 1 is if **M** is a permutation of a diagonal matrix [2]. Equivalently, the only possibility for that is if $\mathbf{x} = \mathbf{s}$ up to permutation and scaling, a typical nonrestrictive indeterminancy in source separation problems.

This effect is illustrated in Fig. 1, which shows a set of three perfectly synchronized sources and their PLFs. That figure also depicts three signals obtained through a linear mixing of the sources, and their PLFs. These mixtures have PLFs lower than 1, in accordance with the result stated in the previous paragraph (even though the PLF between sources 2 and 3 happens to be rather high, but still not 1).

This property illustrates that separation of these sources is necessary to make any type of inference about their synchrony, as measured through the PLF. If they are not properly separated, the synchrony values measured will not be accurate. On the other hand, established BSS methods such as Independent Component Analysis (ICA) are not adequate for this task, since phase-locked sources are not independent [2]. PLMF is a source separation algorithm tailored specifically for this problem, and it is presented in the next section.

3 Algorithm

We begin with a summary of the notation and definitions used in this section; we then formulate the optimization problem for PLMF and present a table of the algorithm at the end.

3.1 Assumptions and General Formulation

We assume that we have a set of N complex-valued sources $s_j(t)$ for $j = 1,\ldots,N$ and $t = 1,\ldots,T$. We assume also that N is known. Denote by $\mathbf{S}$ a N by T complex-valued matrix whose (j,t)th entry is $s_j(t)$. One can easily separate the amplitude and phase components of the sources through $\mathbf{S} = \mathbf{A} \odot \boldsymbol{\Phi}$, where $\odot$ is the elementwise (or Hadamard) product, $\mathbf{A}$ is a real-valued N by T matrix with its (j,t) element defined as $a_j(t) \equiv |s_j(t)|$, and $\boldsymbol{\Phi}$ is a N by T complex-valued matrix with its (j,t) element defined as $\boldsymbol{\Phi}_j(t) \equiv \mathrm{e}^{\mathrm{i}(\mathrm{angle}(s_j(t)))} \equiv \mathrm{e}^{\mathrm{i}\phi_j(t)}$.

The representation of $\mathbf{S}$ in amplitude and phase is, thus far, completely general: it merely represents $\mathbf{S}$ in polar coordinates. We place no constraints on $\mathbf{A}$ other than nonnegativity, since its elements are absolute values of complex numbers. This is consistent with the use of the PLF as a measure of synchrony: the PLF uses no information from the signal amplitudes.

We assume that the sources are perfectly synchronized; as discussed in Sect. 2.2, in this situation, $\Delta\phi_{jk}(t) = \phi_j(t) - \phi_k(t)$ is constant for all t, for any j and k. Thus, $\boldsymbol{\Phi}$ can be decomposed as

$$\boldsymbol{\Phi} \equiv \mathbf{z}\mathbf{f}^{\mathsf{T}}, \tag{2}$$

where $\mathbf{z}$ is a complex-valued column vector of size N containing the relative phase lags of each source, and $\mathbf{f}$ is a complex-valued column vector of size T containing the common oscillation. In simpler terms, if the sources are phase-locked, then $\mathrm{rank}(\boldsymbol{\Phi}) = 1$, and the above decomposition is always possible, even though it is not unique. Then, the time evolution of each source's phase is given by $\phi_j(t) = \mathrm{angle}(z_j) + \mathrm{angle}(f_t)$, where z_j and f_t are the jth entry of $\mathbf{z}$ and the tth element of $\mathbf{f}$, respectively.

Although one can conceive complex-valued sources where the rows of $\mathbf{A}$ and the vector $\mathbf{f}$ vary rapidly with time, in real-world systems we expect them to vary smoothly; for this reason, as will be seen below, we chose to softly enforce the smoothness of these two variables in PLMF.

We also assume that we only have access to P measurements ($P \geq N$) that result from a linear mixing of the sources, as is customary in source separation problems:

$$\mathbf{X} \equiv \mathbf{M}\mathbf{S} + \mathbf{N}, \tag{3}$$

where $\mathbf{X}$ is a P by T matrix containing the measurements, $\mathbf{M}$ is a P by N real-valued mixing matrix and $\mathbf{N}$ is a P by T complex-valued matrix of noise. Our assumption of a real mixing matrix is appropriate in the case of linear and instantaneous mixing, as motivated earlier. We will deal only with the noiseless model, where $\mathbf{N} = 0$, although we then also test how it copes with noisy data.

The goal of PLMF is to recover $\mathbf{S}$ and $\mathbf{M}$ using only $\mathbf{X}$. A simple way to do this is to find $\mathbf{M}$ and $\mathbf{S}$ such that the data misfit, defined as $\frac{1}{2}\left\|\mathbf{X} - \mathbf{M}\left(\mathbf{A} \odot (\mathbf{z}\mathbf{f}^{\mathsf{T}})\right)\right\|_F^2$, where $\|\cdot\|_F$ is the Frobenius norm, is as small as possible. As mentioned above, we also want the estimates of $\mathbf{A}$ and $\mathbf{f}$ to be smooth. Thus, the minimization problem to be solved is given by

$$\min_{\mathbf{M},\mathbf{A},\mathbf{z},\mathbf{f}} \frac{1}{2} \left\| \mathbf{X} - \mathbf{M}\left(\mathbf{A} \odot (\mathbf{z}\mathbf{f}^{\mathsf{T}})\right) \right\|_F^2 + \lambda_{\mathbf{A}} \left\| \mathbf{A}\mathbf{D}_{\mathbf{A}} \right\|_F^2 + \lambda_{\mathbf{f}} \left\| \mathbf{D}_{\mathbf{f}}\mathbf{f} \right\|_2^2, \tag{4}$$

s.t.: 1) All elements of **M** must lie between -1 and $+1$.

2) All elements of **A** must be non-negative.

3) All elements of **z** and **f** must have unit absolute value.

where $\mathbf{D}_{\mathbf{A}}$ and $\mathbf{D}_{\mathbf{f}}$ are the first-order difference operators of appropriate size, such that the entry (j,t) of $\mathbf{A}\mathbf{D}_{\mathbf{A}}$ is given by $a_j(t) - a_j(t+1)$, and the kth entry of $\mathbf{D}_{\mathbf{f}}\mathbf{f}$ is given by $f_k - f_{(k+1)}$. The first term directly measures the misfit between the real data and the product of the estimated mixing matrix and the estimated sources. The second and third terms enforce smoothness of the rows of **A** and of the vector **f**, respectively. These two terms allow for better estimates for **A** and **f** under additive white noise, since enforcing smoothness is likely to filter the high-frequency components of that noise.

Constraint 2 ensures that **A** represents amplitudes, whereas Constraint 3 ensures that **z** and **f** represent phases. Constraint 1 prevents the mixing matrix **M** from exploding to infinity while **A** goes to zero. Note that we also penalize indirectly the opposite indeterminancy, where **M** goes to zero while **A** goes to infinity: that would increase the value of the second term while keeping the other terms constant, as long as the rows of **A** do not have all elements equal to each other. Thus, the solution for **M** lies on the boundary of the feasible set for **M**; using this constraint instead of forcing the L_1 norm of each row to be exactly 1, as was done in [3], makes the subproblem for **M** convex, with all the known advantages that this brings [5].

3.2 Optimization

The minimization problem presented in (4) depends on the four variables **M**, **A**, **z**, and **f**. Although the minimization problem is not globally convex, it is convex in **A** and **M** individually, while keeping the other variables fixed. For simplicity, we chose to optimize (4) in each variable at a time, by first optimizing on **M** while keeping **A**, **z** and **f** constant; then doing the same for **A**, followed by **z**, and then **f**. This cycle is repeated until convergence. From our experience with the method, the particular order in which the variables are optimized is not critical. Although this algorithm is not guaranteed to converge to a global minimum, we have experienced very few cases of local optima.

In the following, we show that the minimization problem above can be translated into well-known forms (constrained least squares problems) for each of the four variables. We also detail the optimization procedure for each of the four subproblems. For brevity, we do not distinguish the real variables such as $\mathbf{M}$ from their estimates $\hat{\mathbf{M}}$ throughout this section: in each subproblem, only one variable is being estimated, while all the others are kept fixed and equal to their current estimates.

Optimization on M If we define $\mathbf{m} \equiv \text{vec}(\mathbf{M})$ and $\mathbf{x} \equiv \text{vec}(\mathbf{X})$,[3] then the minimization subproblem for $\mathbf{M}$, while keeping all other variables fixed, is equivalent to the following constrained least-squares problem:

$$\min_{\mathbf{m}} \frac{1}{2} \left\| \begin{bmatrix} \mathcal{R}(\mathbf{x}) \\ \mathcal{I}(\mathbf{x}) \end{bmatrix} - \begin{bmatrix} \mathcal{R}(\mathbf{R}) \\ \mathcal{I}(\mathbf{R}) \end{bmatrix} \mathbf{m} \right\|_2^2 \tag{5}$$

$$\text{s.t.:} \quad -1 \leq \mathbf{m} \leq +1,$$

where $\mathcal{R}(.)$ and $\mathcal{I}(.)$ are the real and imaginary parts, $\mathbf{I}_P$ is the P by P identity matrix, and $\mathbf{R} \equiv [\mathbf{S}^\mathsf{T} \otimes \mathbf{I}_P]$, with $\otimes$ denoting the Kronecker product and $\|\cdot\|_2$ denoting the Euclidean norm. Here, and throughout this paper, all inequalities should be understood in the componentwise sense, i.e., every entry of $\mathbf{M}$ is constrained to be between -1 and $+1$. For convenience, we used the least-squares solver implemented in the MATLAB Optimization Toolbox to solve this problem, although many other solvers exist.

The main advantage of using the constraint $-1 \leq \mathbf{M} \leq +1$ is now clear: it is very simply translated into $-1 \leq \mathbf{m} \leq +1$ after applying the vec(.) operator, remaining a convex constraint, whereas other constraints would be harder to apply.

Optimization on A The optimization in $\mathbf{A}$ can also be reformulated as a least-squares problem. If $\mathbf{a} \equiv \text{vec}(\mathbf{A})$, the minimization on $\mathbf{A}$ is equivalent to

$$\min_{\mathbf{a}} \frac{1}{2} \left\| \begin{bmatrix} \mathcal{R}(\mathbf{x}) \\ \mathcal{I}(\mathbf{x}) \\ \mathbf{0}_{(N^2-N)} \end{bmatrix} - \begin{bmatrix} \mathcal{R}(\mathbf{K}) \\ \mathcal{I}(\mathbf{K}) \\ \sqrt{\lambda_\mathbf{A}} \mathbf{D}_\mathbf{A} \otimes \mathbf{I}_N \end{bmatrix} \mathbf{a} \right\|_2^2 \tag{6}$$

$$\text{s.t.:} \quad \mathbf{a} \geq 0,$$

where $\mathbf{K} \equiv [(\text{Diag}(\mathbf{f}) \otimes \mathbf{M})\text{Diag}(\mathbf{z}_0)]$, $\mathbf{0}_{(N^2-N)}$ is a column vector of size $(N^2 - N)$, filled with zeros, and Diag(.) is a square diagonal matrix of appropriate dimension having the input vector on the main diagonal. We again use the built-in MATLAB solver to solve this subproblem.

Optimization on z The minimization problem in $\mathbf{z}$ with no constraints is equivalent to:

[3] The vec(.) operator stacks the columns of a matrix into a column vector.

$$\min_{\mathbf{z}} \frac{1}{2} \|\mathbf{Oz} - \mathbf{x}\|_2^2 \quad \text{with} \quad \mathbf{O} = \begin{bmatrix} f_1\mathbf{M}\,\text{Diag}(\mathbf{a}(1)) \\ f_2\mathbf{M}\,\text{Diag}(\mathbf{a}(2)) \\ \vdots \\ f_T\mathbf{M}\,\text{Diag}(\mathbf{a}(T)) \end{bmatrix}, \tag{7}$$

where f_t is the tth entry of $\mathbf{f}$, and $\mathbf{a}(t)$ is the tth column of $\mathbf{A}$. Usually, the solution of this system will not obey the unit absolute value constraint. To circumvent this, we solve this unconstrained linear system and afterwards normalize $\mathbf{z}$ for all sources j and time instants t, by transferring its absolute value onto variable $\mathbf{A}$:

$$a_j(t) \leftarrow |z_j| a_j(t) \quad \text{and} \quad z_j \leftarrow z_j / |z_j|.$$

It is easy to see that the new $\mathbf{z}$ obtained after this normalization is still a global minimizer of (7) (where the new value of $\mathbf{A}$ should be used).

Optimization on f Let $\tilde{\mathbf{x}} \equiv \text{vec}(\mathbf{X}^\mathsf{T})$. The minimization problem in $\mathbf{f}$ with no constraints can be shown to be equivalent to

$$\min_{\mathbf{f}} \frac{1}{2} \left\| \begin{bmatrix} \mathbf{P} \\ \sqrt{\lambda_\mathbf{f}}\mathbf{D}_\mathbf{f} \end{bmatrix} \mathbf{f} - \begin{bmatrix} \tilde{\mathbf{x}} \\ \mathbf{0}_{(N-1)} \end{bmatrix} \right\|_2^2 \tag{8}$$

$$\text{with} \quad \mathbf{P} = \begin{bmatrix} \sum_j m_{1j} z_j \text{Diag}(\mathbf{a}_j) \\ \sum_j m_{2j} z_j \text{Diag}(\mathbf{a}_j) \\ \vdots \\ \sum_j m_{Pj} z_j \text{Diag}(\mathbf{a}_j) \end{bmatrix},$$

where f_t is the tth entry of $\mathbf{f}$, m_{ij} is the (i,j) entry of $\mathbf{M}$, z_j is the jth entry of $\mathbf{z}$, and $\mathbf{a}_j$ is the jth row of $\mathbf{A}$.

As in the subproblem for $\mathbf{z}$, in general the solution of this system will not obey the unit absolute value constraint. Thus, we perform a similar normalization, given by

$$a_j(t) \leftarrow |f_t| a_j(t) \quad \text{and} \quad f_t \leftarrow f_t / |f_t|. \tag{9}$$

Note that unlike the previous case of the optimization for $\mathbf{z}$, this normalization changes the cost function, in particular the term $\lambda_\mathbf{f} \|\mathbf{D}_\mathbf{f}\mathbf{f}\|_2^2$. Therefore, there is no guarantee that after this normalization we have found a global minimum for $\mathbf{f}$.

For this reason, we construct a vector of angles $\beta \equiv \text{angle}(\mathbf{f})$ and minimize the cost function (4) as a function of β, using 20 iterations of Newton's algorithm. Although infinitely many values of β correspond to a given $\mathbf{f}$, any of those values is suitable. The advantage of using this new variable is that there are no constraints in β, so the Newton algorithm can be used freely. Thus, the normalized solution of the linear system in (8) can be considered simply as an initialization for the Newton algorithm on β, which in most conditions can find a local minimum.

On the first time that the Newton algorithm is run, it is initialized using the unconstrained problem (8) and the ensuing normalization (9). On the second and following times that it is run, the result of the previous minimization on $\mathbf{f}$ is used as initial value.

Phase-Locked Matrix Factorization The consecutive cycling of optimizations on $\mathbf{M}$, $\mathbf{A}$, $\mathbf{z}$ and $\mathbf{f}$ constitutes the Phase-Locked Matrix Factorization (PLMF) algorithm. A summary of this algorithm is presented below.

	PHASE-LOCKED MATRIX FACTORIZATION
1:	Input data $\mathbf{X}$
2:	Input random initializations $\hat{\mathbf{M}}$, $\hat{\mathbf{A}}$, $\hat{\mathbf{z}}$, $\hat{\mathbf{f}}$
3:	**for** iter $\in \{1,2,\ldots,\text{MaxIter}\}$, **do**
4:	Solve the constrained problem in Eq. (3.2)
5:	Solve the constrained problem in Eq. (6)
6:	Solve the unconstrained system in Eq. (7)
7:	$a_j(t) \leftarrow \lvert z_j \rvert a_j(t)$ and $z_j \leftarrow z_j / \lvert z_j \rvert$, $j = 1, \ldots, N$
8:	**if** iter = 1
9:	Solve the unconstrained system in Eq. (8)
10:	$a_j(t) \leftarrow \lvert f_t \rvert a_j(t)$ and $f_t \leftarrow f_t / \lvert f_t \rvert$, $t = 1, \ldots, T$
11:	Optimize $\beta \equiv \text{angle}(\mathbf{f})$ with Newton algorithm (use result of step 10 as initialization)
12:	**else**
13:	Optimize $\beta \equiv \text{angle}(\mathbf{f})$ with Newton algorithm (use Newton algorithm from (iter-1) as init.)
14:	**end for**

4 Simulation and Results

In this section we show results on small simulated datasets, demonstrating that PLMF can correctly factor the data $\mathbf{X}$ into a mixing matrix $\mathbf{M}$, amplitudes $\mathbf{A}$, and phases $\mathbf{z}$ and $\mathbf{f}$. Despite deriving PLMF for the noiseless case, we will also test its robustness to a small noisy perturbation.

4.1 Data Generation

We generate the data directly from the model $\mathbf{X} = \mathbf{MS}$, with $\mathbf{S} = \mathbf{A} \odot \mathbf{\Phi} = \mathbf{A} \odot (\mathbf{z}\mathbf{f}^{\mathsf{T}})$, taking $N = 2$ and $P = 4$. The number of time samples is $T = 100$. $\mathbf{M}$ is taken as a random matrix with entries uniformly distributed between -1 and $+1$. We then normalize $\mathbf{M}$ so that the entry with the largest absolute value is ± 1. Each row of $\mathbf{A}$ (i.e. each source's amplitude) is generated as a sum of a constant baseline and 2–5

Gaussians with random mean and random variance. $\mathbf{z}$ is always equal to $\left[0, \frac{2\pi}{3}\right]^{\mathsf{T}}$.[4] $\mathbf{f}$ is generated as a complex sinusoid with angular frequency 0.06 in the first half of the observation period, and angular frequency 0.04 in its second half, in a way that $\mathbf{f}$ has no discontinuities. $\mathbf{X}$ is then generated according to the data model: $\mathbf{X} = \mathbf{M}(\mathbf{A} \odot (\mathbf{z}\mathbf{f}^{\mathsf{T}}))$.

The initial values for the estimated variables are all random: elements of $\hat{\mathbf{M}}$ and $\hat{\mathbf{A}}$ are drawn from the Uniform([0,1]) distribution ($\hat{\mathbf{M}}$ is then normalized in the same way as $\mathbf{M}$), while the elements of $\hat{\mathbf{z}}$ are of the form $e^{i\alpha}$ with α taken from the Uniform $\left(\left[0, \frac{\pi}{2}\right]\right)$ distribution. The elements of $\mathbf{f}$ are also of the form $e^{i\beta}$, with β uniformly distributed between 0 and 2π.

We generate 100 datasets of two types with the following features:

- 100 noiseless datasets: 2 sources, 4 sensors, 100 time points, no noise.
- 100 noisy datasets: same as above 1, but with added complex Gaussian white noise as in (3). The noise power is such that the Signal-to-Noise Ratio (SNR) of the data is 20 dB.

4.2 *Quality Measures*

$\hat{\mathbf{M}}$ can be compared with $\mathbf{M}$ through the gain matrix $\mathbf{G} \equiv \hat{\mathbf{M}}^{+}\mathbf{M}$, where $\hat{\mathbf{M}}^{+}$ is the Moore-Penrose pseudo-inverse of $\hat{\mathbf{M}}$ [4]. This is the same as $\hat{\mathbf{M}}^{-1}\mathbf{M}$ if the number of sensors is equal to the number of sources. If the estimation is well done, the gain matrix should be close to a permutation of the identity matrix. After manually compensating a possible permutation of the estimated sources, we compare the sum of the squares of the diagonal elements of $\mathbf{G}$ with the sum of the squares of its off-diagonal elements. This criterion is called Signal-to-Interference Ratio (SIR).

Also, $\hat{\mathbf{A}}$ will be compared to $\mathbf{A}$ through visual inspection for one dataset with an SIR close to the average SIR of the 100 datasets.

4.3 *Results*

We did not implement a convergence criterion; we simply do 400 cycles of the optimization on $\mathbf{M}$, $\mathbf{A}$, $\mathbf{z}$ and $\mathbf{f}$ using $\lambda_{\mathbf{A}} = 3$ and $\lambda_{\mathbf{f}} = 1$. The mean and standard deviation of the SIR criterion are presented in Table 1. This table also shows results for other choices of $\lambda_{\mathbf{A}}$, which are discussed in Sect. 5.1. Figure 2 shows the results of the estimation of the source amplitudes for one representative dataset, showing that $\hat{\mathbf{A}}$ is quite close to the real $\mathbf{A}$ for both the noiseless and the noisy datasets. Note

[4]This choice of $\mathbf{z}$ is done to ensure that the sources never have phase lags close to 0 or π, which violate the mild assumptions mentioned in Sect. 2.3 [2].

Table 1 Comparison of the estimated mixing matrix $\hat{\mathbf{M}}$ with the true mixing matrix $\mathbf{M}$ through the pseudo-SIR of the gain matrix $\mathbf{G} \equiv \hat{\mathbf{M}}^{+}\mathbf{M}$

	SIR (dB)		
Data	$\lambda_{\mathbf{A}} = 0.3$	$\lambda_{\mathbf{A}} = 3$	$\lambda_{\mathbf{A}} = 30$
Dataset 1	23.93 ± 12.72	24.67 ± 12.27	17.34 ± 10.46
Dataset 2	24.37 ± 13.02	25.20 ± 12.29	19.33 ± 10.96

For zero noise (dataset 1), the estimation is quite good, and the performance hit due to the presence of noise (dataset 2) is minimal. We used $\lambda_{\mathbf{f}} = 1$ for all the entries of the table.

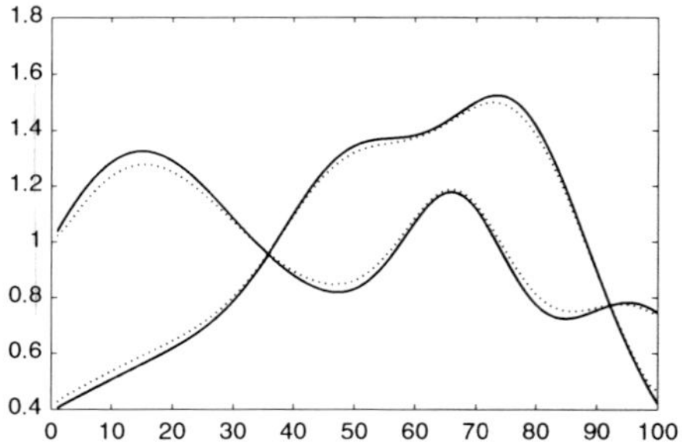

Fig. 2 Visual comparison of the estimated amplitudes $\hat{\mathbf{A}}$ (*dashed lines*) with the true amplitudes $\mathbf{A}$ (*solid lines*), for a representative dataset, after both are normalized so that they have unit means over the observation period. (*Left*) Results for a noiseless dataset: the two estimated amplitudes are close to the true values. Note that the estimated amplitudes have slower variations than the true amplitudes, due to the term with $\lambda_{\mathbf{A}}$. (*Right*) Results for a noisy dataset: due to the presence of noise, it is impossible for the two estimated amplitudes to coincide perfectly with the true ones, but nevertheless the estimated amplitudes follow the real ones very closely. For this figure, the values $\lambda_{\mathbf{A}} = 3$ and $\lambda_{\mathbf{f}} = 1$ were used

that if noise is present, it is impossible to recreate the original amplitudes as they are only present in the data corrupted by noise: one can thus only estimate the corrupted amplitudes. If desired, a simple low-pass filtering procedure can closely recreate the original amplitudes.

These results illustrate that PLMF can separate phase-locked sources in both the noiseless and the noisy condition. Furthermore, they show that there is no performance hit due to the presence of a small amount of noise, suggesting that PLMF has good robustness against small perturbations.

5 Discussion

The above results show that this approach has a high potential, although some limitations must be addressed to turn this algorithm practical for real-world applications.

One incomplete aspect of PLMF is its lack of a stopping criterion; in fact, the results shown in Table 1 could be considerably improved if the number of iterations is increased to, say, 1,000, although that is not the case for all of the 100 datasets. We did not tackle this aspect due to lack of time; however, the data misfit (first term of the cost function) can probably be used to design a decent criterion.

If the sources are not perfectly phase-locked, their pairwise phase differences $\Delta\phi_{ij}$ are not constant in time and therefore one cannot represent the source phases by a single vector of phase lags $\mathbf{z}$ and a single vector $\mathbf{f}$ with a common oscillation. In other words, $\mathbf{\Phi}$ will have a rank higher than 1 (in most cases, it will have the maximum possible rank, which is N), which makes a representation $\mathbf{\Phi} = \mathbf{z}\mathbf{f}^{\mathsf{T}}$ impossible. We are investigating a way to estimate the "most common" phase oscillation $\mathbf{f}$ from the data $\mathbf{X}$, after which PLMF can be used to initialize a more general algorithm that estimates the full $\mathbf{\Phi}$. We are currently testing also a more general algorithm, which optimizes $\mathbf{\Phi}$ with a gradient descent algorithm. Yet, it is somewhat prone to local minima, as one would expect for optimizing variables of size $NT = 200$. A good initialization is likely to alleviate this problem.

Another limitation of PLMF is the indetermination that arises if two sources have $\Delta\phi_{ij} = 0$ or π. In that case, the problem becomes ill-posed, as was already the case in IPA [2]. In fact, using sources with $\Delta\phi_{ij} < \frac{\pi}{10}$ starts to deteriorate the results of PLMF, even with zero noise.

One further aspect which warrants discussion is PLMF's identifiability. If we find two factorizations such that $\mathbf{X} = \mathbf{M}_1\left(\mathbf{A}_1 \odot (\mathbf{z}_1\mathbf{f}_1^{\mathsf{T}})\right) = \mathbf{M}_2\left(\mathbf{A}_2 \odot (\mathbf{z}_2\mathbf{f}_2^{\mathsf{T}})\right)$ (i.e., two factorizations which perfectly describe the same data $\mathbf{X}$), does that imply that $\mathbf{M}_1 = \mathbf{M}_2$, and similar equalities for the other variables? It is quite clear that the answer is negative: the usual indeterminancies of BSS apply to PLMF as well, namely the indeterminancies of permutation, scaling, and sign of the estimated sources. There is at least one further indeterminancy: starting from a given solution $\mathbf{X} = \mathbf{M}_1\left(\mathbf{A}_1 \odot (\mathbf{z}_1\mathbf{f}_1^{\mathsf{T}})\right)$, one can always construct a new one by defining $\mathbf{z}_2 \equiv e^{i\psi}\mathbf{z}_1$ and $\mathbf{f}_2 \equiv e^{-i\psi}\mathbf{f}_1$, while keeping $\mathbf{M}_2 \equiv \mathbf{M}_1$ and $\mathbf{A}_2 \equiv \mathbf{A}_1$. Note that $\mathbf{S}_1 = \mathbf{A}_1 \odot (\mathbf{z}_1\mathbf{f}_1^{\mathsf{T}}) = \mathbf{A}_2 \odot (\mathbf{z}_2\mathbf{f}_2^{\mathsf{T}}) = \mathbf{S}_2$, thus the estimated sources are exactly the same.

5.1 Choice of Parameters $\lambda_{\mathbf{A}}$ and $\lambda_{\mathbf{f}}$

The values of the parameters that we chose were somewhat ad hoc. However, PLMF is rather robust to the choice of $\lambda_{\mathbf{A}}$. Table 1 shows not only the values for $\lambda_{\mathbf{A}} = 3$ (and $\lambda_{\mathbf{f}} = 1$), but also for cases where $\lambda_{\mathbf{A}}$ is one order of magnitude smaller (0.3) or greater (30). Those results show that the SIR does not change too much when this parameter varies by two orders of magnitude.

$\lambda_{\mathbf{A}}$ has the effect of penalizing large variations in $\mathbf{A}$ which can be due to the presence of noise. Therefore, if this parameter is too large, the algorithm will underestimate the variations present in the true amplitudes, as illustrated in Fig. 3. In this figure, the shape of the estimated amplitudes is similar to the shape of the true amplitudes, but the variations are smaller. This effect was already present in Fig. 2 for $\lambda_{\mathbf{A}} = 3$, but in that case the error was very slight, whereas in Fig. 3 the effect is

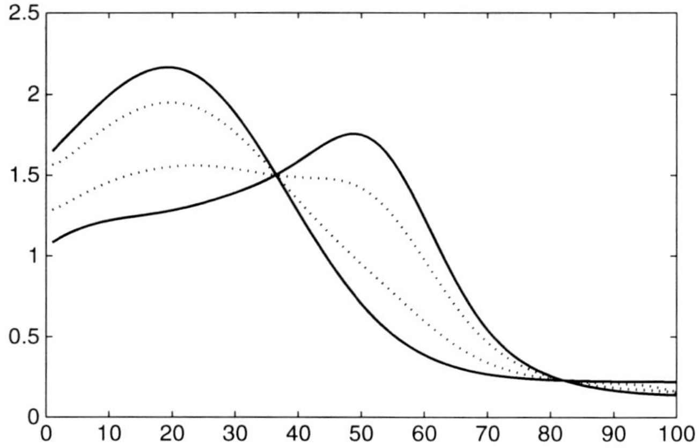

Fig. 3 A typical result of choosing a value of $\lambda_\mathbf{A}$ which is too large (in this case, $\lambda_\mathbf{A} = 30$). The true amplitudes are shown in *solid lines*, whereas estimated amplitudes are shown in *dashed lines*. The estimated amplitudes are similar in shape to the true ones, but have lower variations, since the penalty term is too strong

very noticeable. This is the reason why $\lambda_\mathbf{A} = 30$ yields somewhat lower SIR values when compared to $\lambda_\mathbf{A} = 3$: the algorithm will yield estimated amplitudes which are smaller than the true amplitudes.

One might then think that the correct way to choose $\lambda_\mathbf{A}$ would be to pick the smallest possible value, which is $\lambda_\mathbf{A} = 0$. The results for $\lambda_\mathbf{A} = 0.3$, from Table 1, might encourage that decision. However, $\lambda_\mathbf{A} = 0$ is a poor choice, because the $\lambda_\mathbf{A}$ term has the indirect effect of preventing $\mathbf{A}$ from exploding to infinity and $\mathbf{M}$ from shrinking to zero, while keeping the product $\mathbf{X} = \mathbf{M}(\mathbf{A} \odot (\mathbf{z}\mathbf{f}^\mathsf{T}))$ constant, a situation which causes severe numerical problems.[5] Therefore, one should pick a small but positive value for $\lambda_\mathbf{A}$.

The effect of $\lambda_\mathbf{f}$ is easier to understand, because the indirect effect mentioned in the last paragraph is not present. This parameter has a much smaller effect if $\mathbf{f}$ is smooth, as is the case studied in this paper. A nonzero value helps with numerical conditioning of the problem in the presence of noise, because it prevents the fast variations of the noise from contaminating the estimated $\mathbf{f}$. However, in contrast to the poor choice $\lambda_\mathbf{A} = 0$ discussed in the previous paragraph, $\lambda_\mathbf{f} = 0$ is a perfectly valid choice.

[5]These numerical problems are the reason why no results for $\lambda_\mathbf{A} = 0$ are shown in this paper.

6 Conclusion

We presented an improved version of Phase-Locked Matrix Factorization (PLMF), an algorithm that directly tries to reconstruct a set of measured signals as a linear mixing of phase-locked sources, by factorizing the data into a product of four variables: the mixing matrix, the source amplitudes, their phase lags, and a common oscillation.

PLMF is now able to estimate the sources even when their common oscillation is unknown—an advantage which greatly increases the applicability of the algorithm. Furthermore, the subproblem for $\mathbf{M}$ is now convex, and the subproblems for $\mathbf{z}$ and $\mathbf{f}$ are tackled in a more appropriate manner which should find local minima. The results show good performance for the noiseless case and good robustness to small amounts of noise. The results show as well that the proposed algorithm is accurate and can deal with low amounts of noise, under the assumption that the sources are fully phase-locked, even if the common oscillation is unknown. This generalization brings us considerably closer to being able to solve the Separation of Synchronous Sources (SSS) problem in real-world data.

Acknowledgements This work was partially funded by the DECA-Bio project of the Institute of Telecommunications, and by the Academy of Finland through its Centres of Excellence Program 2006–2011.

References

1. Almeida, M., Schleimer, J.-H., Vigário, R. V., Dias, J.: "Source Separation and Clustering of Phase-Locked Subspaces", IEEE Trans. on Neural networks, **22**(9), pp. 1419–1434 (2011)
2. Almeida, M., Schleimer, J.-H., Bioucas-Dias, J., Vigário, R.: Source separation and clustering of phase-locked subspaces. IEEE Trans. Neural Networks **22**(9), 1419–1434 (2011)
3. Almeida, M., Vigário, R.V., Dias, J.: "Phase Locked Matrix Factorization", Proc European Signal Processing Conf. - EUSIPCO, Barcelona, Spain, **0**, 1728–1732, (2011)
4. Ben-Israel, A., Greville, T.: Generalized Inverses: Theory and Applications. Springer, Berlin (2003)
5. Boyd, S., Vandenberghe, L.: Convex Optimization. Cambridge University Press, Cambridge (2004)
6. Gold, B., Oppenheim, A.V., Rader, C.M.: Theory and implementation of the discrete hilbert transform. In: Rabiner, L.R., Rader, C.M. (eds.) Discrete Signal Processing, John Wiley & Sons Inc; 1st edition, (1973)
7. Hyvärinen, A., Karhunen, J., Oja, E.: Independent Component Analysis. Wiley, New York (2001)
8. Todd K. Leen, Thomas G. Dietterich, Volker Tresp In Advances in Neural Information Processing Systems 13, pp. 556–562 (2001)
9. Nunez, P.L., Srinivasan, R., Westdorp, A.F., Wijesinghe, R.S., Tucker, D.M., Silberstein, R.B., Cadusch, P.J.: EEG coherency I: statistics, reference electrode, volume conduction, laplacians, cortical imaging, and interpretation at multiple scales. Electroencephalogr. Clin. Neurophysiol. **103**, 499–515 (1997)

10. Palva, J.M., Palva, S., Kaila, K.: Phase synchrony among neuronal oscillations in the human cortex. J. Neurosci. **25**(15), 3962–3972 (2005)
11. Pikovsky, A., Rosenblum, M., Kurths, J.: In: Synchronization: A Universal Concept in Nonlinear Sciences. Cambridge Nonlinear Science Series. Cambridge University Press, Cambridge (2001)
12. Schoffelen, J.-M., Oostenveld, R., Fries, P.: Imaging the human motor system's beta-band synchronization during isometric contraction. NeuroImage **41**, 437–447 (2008)
13. Torrence, C., Compo, G.P.: A practical guide to wavelet analysis. Bull. Am. Meteorol. Soc. **79**, 61–78 (1998)
14. Uhlhaas, P.J., Singer, W.: Neural synchrony in brain disorders: Relevance for cognitive dysfunctions and pathophysiology. Neuron **52**, 155–168 (2006)
15. Vigário, R., Särelä, J., Jousmäki, V., Hämäläinen, M., Oja, E.: Independent component approach to the analysis of EEG and MEG recordings. IEEE Trans. Biom. Eng. **47**(5), 589–593 (2000)
16. Niklasson, L., Bodén, M., Ziemke, T.: Perspectives in Neural Computing, Proceedings of the 8th International Conference on Artificial Neural Networks, ICANN'98, Springer Verlag, 675–680 (1998)

Stochastic Subgradient Estimation Training for Support Vector Machines

Sangkyun Lee and Stephen J. Wright

Abstract Subgradient algorithms for training support vector machines have been successful in solving many large-scale and online learning problems. However, for the most part, their applicability has been restricted to linear kernels and strongly convex formulations. This paper describes efficient subgradient approaches without such limitations. Our approaches make use of randomized low-dimensional approximations to nonlinear kernels, and minimization of a reduced primal formulation using an algorithm based on robust stochastic approximation, which does not require strong convexity. Experiments illustrate that our approaches produce solutions of comparable prediction accuracy with the solutions acquired from existing SVM solvers, but often in much shorter time.

1 Introduction

Support vector machines (SVMs) have been highly successful in machine learning and data mining. We broadly categorize training algorithms for SVMs as follows:

1. *Decomposition methods* based on dual formulations, such as SVM-Light [8], LIBSVM [6], and an online variant LASVM [1]. Kernels can be easily introduced into formulations via the kernel trick [2].

S. Lee (✉)
Fakultät für Informatik, LS VIII, Technische Universität Dortmund, 44221 Dortmund, Germany
e-mail: sangkyun.lee@tu-dortmund.de

S.J. Wright
Computer Sciences Department, University of Wisconsin, 1210 W. Dayton Street, Madison, WI 53706, USA
e-mail: swright@cs.wisc.edu

P. Latorre Carmona et al. (eds.), *Mathematical Methodologies in Pattern Recognition and Machine Learning*, Springer Proceedings in Mathematics & Statistics 30,
DOI 10.1007/978-1-4614-5076-4_5,

2. *Cutting-plane methods* using special primal formulations to successively add violated constraints. SVM-Perf [9] and OCAS [7] handle linear kernels; the former approach is extended for nonlinear kernels in CPNY [11] and CPSP [10].
3. *Subgradient methods* for primal formulations with linear kernels, including Pegasos [18] and SGD [3]. The former is adapted in [19] to nonlinear kernels.

Subgradient methods are of particular interest in this paper since they are well suited to large-scale and online learning problems.

This paper aims to provide practical subgradient algorithms for training SVMs with nonlinear kernels, overcoming the weakness of the updated Pegasos algorithm [19], which uses exact kernel information and which may (in the worst case) require a dual variable for each training example. Our approach uses a primal formulation with low-dimensional approximations to feature mappings. Such approximations are obtained either by approximating the Gram matrix or by constructing subspaces whose random bases approximate the feature spaces induced by kernels. These approximations can be computed and applied to data points iteratively and thus are suited to an online context as well. Further, we suggest an efficient way to make predictions for test points using the approximate feature mappings, without recovering the potentially large number of support vectors.

Unlike Pegasos, we use Vapnik's original SVM formulation without modifying the objective to be strongly convex. Our main algorithm takes steplengths of size $O(1/\sqrt{t})$ (associated with the robust stochastic approximation methods [14, 15] and online convex programming [20]), rather than the $O(1/t)$ steplength scheme in Pegasos. As we show in the experiments, there is little practical difference between $O(1/\sqrt{t})$ steplengths and $O(1/t)$ steps.

2 Nonlinear SVMs in the Primal

In this section we discuss the primal SVM formulation in a low-dimensional feature space induced by kernel approximation.

2.1 *Structure of the Formulation*

We first analyze the structure of the primal SVM formulation with nonlinear feature mappings. To unveil the details, we apply the tools of convex analysis, rather than appealing to the representer theorem [12] as in [4].

Let us consider the training point and label pairs $\{(\mathbf{t}_i, y_i)\}_{i=1}^m$ for $\mathbf{t}_i \in \mathbb{R}^n$ and $y_i \in \mathbb{R}$, and a feature mapping $\phi : \mathbb{R}^n \to \mathbb{R}^d$. Given a convex loss function $\ell(\cdot) : \mathbb{R} \to \mathbb{R} \cup \{\infty\}$ and $\lambda > 0$, the primal SVM problem (for classification) can be stated as follows:

$$\text{(P1)} \quad \min_{\mathbf{w}\in\mathbb{R}^d, b\in\mathbb{R}} \frac{\lambda}{2}\mathbf{w}^{\mathrm{T}}\mathbf{w} + \frac{1}{m}\sum_{i=1}^{m} \ell\left(y_i\left(\mathbf{w}^{\mathrm{T}}\phi\left(\mathbf{t}_i\right) + b\right)\right).$$

The necessary and sufficient optimality conditions are

$$\lambda \mathbf{w} + \frac{1}{m}\sum_{i=1}^{m} \chi_i y_i \phi(\mathbf{t}_i) = 0,\ \frac{1}{m}\sum_{i=1}^{m} \chi_i y_i = 0 \text{ for } \chi_i \in \partial\ell\left(y_i\left(\mathbf{w}^{\mathrm{T}}\phi\left(\mathbf{t}_i\right) + b\right)\right), \tag{1}$$

where $\partial\ell$ is the subdifferential of ℓ. We now consider the following substitution:

$$\mathbf{w} = \sum_{i=1}^{m} \alpha_i \phi(\mathbf{t}_i) \tag{2}$$

(which mimics the form of the first equality in (1)). Motivated by this expression, we formulate the following problem

$$\text{(P2)} \quad \min_{\alpha\in\mathbb{R}^m, b\in\mathbb{R}} \frac{\lambda}{2}\alpha^{\mathrm{T}}\Psi\alpha + \frac{1}{m}\sum_{i=1}^{m} \ell\left(y_i(\Psi_{i\cdot}\alpha + b)\right),$$

where $\Psi \in \mathbb{R}^{m\times m}$ is defined by $\Psi_{ij} := \phi(\mathbf{t}_i)^{\mathrm{T}}\phi(\mathbf{t}_j)$ for $i, j = 1, 2, \ldots, m$, and $\Psi_{i\cdot}$ denotes the ith row of Ψ. Optimality conditions for (P2) are as follows:

$$\lambda\Psi\alpha + \frac{1}{m}\sum_{i=1}^{m} \beta_i y_i \Psi_{i\cdot}^{\mathrm{T}} = 0,\ \frac{1}{m}\sum_{i=1}^{m} \beta_i y_i = 0 \text{ for } \beta_i \in \partial\ell\left(y_i(\Psi_{i\cdot}\alpha + b)\right). \tag{3}$$

In the following result, we show that the solution of (P1) can be derived from a solution of (P2). This result can be regarded as a specialized representer theorem.

Proposition 1. *Let* $(\alpha, b) \in \mathbb{R}^m \times \mathbb{R}$ *be a solution of (P2). Then if we define* $\mathbf{w}$ *by (2),* $(\mathbf{w}, b) \in \mathbb{R}^d \times \mathbb{R}$ *is a solution of (P1).*

Proof. Since (α, b) solves (P2), (3) hold for some β_i, $i = 1, \ldots, m$. To prove the claim, it suffices to show that $(\mathbf{w}, b)$ and χ satisfy (1), where $\mathbf{w}$ is defined by (2) and $\chi_i = \beta_i$ for all $i = 1, \ldots, m$. Substituting $\Psi_{ij} = \phi(\mathbf{t}_i)^{\mathrm{T}}\phi(\mathbf{t}_j)$ in (3), we have

$$\lambda\sum_{i=1}^{m} \phi(\mathbf{t}_j)^{\mathrm{T}}\phi(\mathbf{t}_i)\alpha_i + \frac{1}{m}\sum_{i=1}^{m} \beta_i y_i \phi(\mathbf{t}_j)^{\mathrm{T}}\phi(\mathbf{t}_i) = 0,\ \frac{1}{m}\sum_{i=1}^{m} \beta_i y_i = 0,$$

where $\beta_i \in \partial\ell\left(y_i\left(\sum_{j=1}^{m} \phi(\mathbf{t}_j)^{\mathrm{T}}\phi(\mathbf{t}_i)\alpha_j + b\right)\right)$. From the first equality, we have

$$-\sum_{i=1}^{m}\left(\alpha_i + \frac{y_i}{\lambda m}\beta_i\right)\phi(\mathbf{t}_i) + \xi = 0, \text{ for } \xi \in \mathrm{Null}\left(\left[\phi(\mathbf{t}_j)^{\mathrm{T}}\right]_{j=1}^{m}\right).$$

Since the two components in this sum are orthogonal, it implies that

$$0 = \left\|\sum_{i=1}^{m}\left(\alpha_i + \frac{y_i}{\lambda m}\beta_i\right)\phi(\mathbf{t}_i)\right\|_2^2 + \xi^{\mathrm{T}}\xi \ \Rightarrow\ \xi = 0.$$

We can therefore rewrite the optimality conditions for (P2) as follows:

$$\sum_{i=1}^{m}\left(\lambda\alpha_i + \frac{y_i}{m}\beta_i\right)\phi(\mathbf{t}_i) = 0,\; \frac{1}{m}\sum_{i=1}^{m}\beta_i y_i = 0, \tag{4}$$

for some $\beta_i \in \partial\ell\left(y_i\left(\phi(\mathbf{t}_i)^{\mathrm{T}}\sum_{j=1}^{m}\alpha_j\phi(\mathbf{t}_j)+b\right)\right)$. By defining $\mathbf{w}$ as in (2) and setting $\chi_i = \beta_i$ for all i, we see that (4) is identical to (1), as claimed. □

While Ψ is clearly symmetric positive semidefinite, the proof makes no assumption about nonsingularity of this matrix, or uniqueness of the solution α of (P2). However, the first equality in (3) suggests that without loss of generality, we can constrain α to have the form $\alpha_i = -\frac{y_i}{\lambda m}\beta_i$, where $\beta_i \in \partial\ell$. (For the hinge loss function $\ell(\delta) := \max\{1-\delta, 0\}$, we have $\beta_i \in [-1,0]$.) These results clarify the connection between the expansion coefficient α and the dual variable $\beta(=\chi)$, which is introduced in [4] but not fully explicated there. Similar arguments for the regression with the ϵ-insensitive loss function $\ell'(\delta) := \max\{|\delta|-\epsilon, 0\}$ lead to $\alpha'_i = -\frac{1}{\lambda m}\beta'_i$, where $\beta'_i \in [-1,1]$ is in $\partial\ell'$.

2.2 *Reformulation Using Approximations*

Consider the original feature mapping $\phi^\circ : \mathbb{R}^n \to \mathcal{H}$ to a Hilbert space $\mathcal{H}$ induced by a kernel $k^\circ : \mathbb{R}^n \times \mathbb{R}^n \to \mathbb{R}$, where k° satisfies the conditions of Mercer's Theorem [17]. Suppose that we have a low-dimensional approximation $\phi : \mathbb{R}^n \to \mathbb{R}^d$ of ϕ° for which

$$k^\circ(\mathbf{s},\mathbf{t}) \approx \phi(\mathbf{s})^{\mathrm{T}}\phi(\mathbf{t}), \tag{5}$$

for all inputs $\mathbf{s}$ and $\mathbf{t}$ of interest. If we construct a matrix $V \in \mathbb{R}^{m\times d}$ by defining the ith row as

$$V_{i\cdot} = \phi(\mathbf{t}_i)^{\mathrm{T}},\; i = 1,2,\ldots,m, \tag{6}$$

then we have

$$\Psi := VV^{\mathrm{T}} \approx \Psi^\circ := [k^\circ(\mathbf{t}_i,\mathbf{t}_j)]_{i,j=1,2,\ldots,m}. \tag{7}$$

Note that Ψ is a positive semidefinite rank-d approximation to Ψ°. Substituting $\Psi = VV^{\mathrm{T}}$ and $\gamma = V^{\mathrm{T}}\alpha$ in (P2) leads to the equivalent formulation

$$\text{(PL)} \quad \min_{\gamma\in\mathbb{R}^d,\, b\in\mathbb{R}} \; \frac{\lambda}{2}\gamma^{\mathrm{T}}\gamma + \frac{1}{m}\sum_{i=1}^{m}\ell(y_i(V_{i\cdot}\gamma+b)).$$

This problem can be regarded as a *linear* SVM with transformed feature vectors $V_{i\cdot}^{\mathrm{T}} \in \mathbb{R}^d$, $i = 1,2,\ldots,m$. An approximate solution to (PL) can be obtained with the subgradient algorithms discussed later in Sect. 3.

2.3 Approximating the Kernel

We discuss two techniques for finding V that satisfies (7). The first uses randomized linear algebra to calculate a low-rank approximation to the matrix Ψ°. The second approach uses random projections to construct approximate feature mappings ϕ explicitly.

Kernel Matrix Approximation In this approach, we specify some integers d and s with $0 < d \le s < m$, and choose s elements at random from the index set $\{1,2,\ldots,m\}$ to form a subset $\mathcal{S}$. We then find the best rank-d approximation $W_{\mathcal{S},d}$ to $(\Psi^\circ)_{\mathcal{S}\mathcal{S}}$, and its pseudo-inverse $W^+_{\mathcal{S},d}$. We choose V so that

$$VV^{\mathrm{T}} = (\Psi^\circ)_{\cdot\mathcal{S}} W^+_{\mathcal{S},d} (\Psi^\circ)^{\mathrm{T}}_{\cdot\mathcal{S}}, \tag{8}$$

where $(\Psi^\circ)_{\cdot\mathcal{S}}$ denotes the column submatrix of Ψ° defined by the indices in $\mathcal{S}$. The results in [5] indicate that in expectation and with high probability, the rank-d approximation obtained by this process has an error that can be made as close to the *best* rank-d approximation by choosing s sufficiently large.

To obtain $W_{\mathcal{S},d}$, we form the eigen-decomposition $(\Psi^\circ)_{\mathcal{S}\mathcal{S}} = QDQ^{\mathrm{T}}$, where $Q \in \mathbb{R}^{s\times s}$ is orthogonal and D is a diagonal matrix with nonincreasing nonnegative diagonal entries. Taking $\bar{d} \le d$ to be the number of positive diagonals in D, we have $W_{\mathcal{S},d}$ and its pseudo-inverse $W^+_{\mathcal{S},d}$ as

$$W_{\mathcal{S},d} = Q_{\cdot,1..\bar{d}} D_{1..\bar{d},1..\bar{d}} Q^{\mathrm{T}}_{\cdot,1..\bar{d}}, \quad W^+_{\mathcal{S},d} = Q_{\cdot,1..\bar{d}} D^{-1}_{1..\bar{d},1..\bar{d}} Q^{\mathrm{T}}_{\cdot,1..\bar{d}},$$

(where $Q_{\cdot,1..\bar{d}}$ denotes the first $\bar{d}$ columns of Q, and so on). Therefore the matrix V satisfying (8) is

$$V = (\Psi^\circ)_{\cdot\mathcal{S}} Q_{\cdot,1..\bar{d}} D^{-1/2}_{1..\bar{d},1..\bar{d}}. \tag{9}$$

For practical implementation, rather than defining d a priori, we can choose a threshold ϵ_d with $0 < \epsilon_d \ll 1$, then choose d to be the largest integer in $1,2,\ldots,s$ such that $D_{dd} \ge \epsilon_d$. (In this case, we have $\bar{d} = d$.)

For each sample set $\mathcal{S}$, this approach requires $O(ns^2 + s^3)$ operations for the creation and factorization of $(\Psi^\circ)_{\mathcal{S}\mathcal{S}}$, assuming the evaluation of each kernel entry takes $O(n)$. Since our algorithm requires a single row of V in each iteration, the computation cost of (9) can be amortized over iterations: the cost is $O(sd)$ per iteration if the corresponding row of Ψ° is available; $O(ns + sd)$ otherwise.

Feature Mapping Approximation The second approach to defining V finds a mapping $\phi : \mathbb{R}^n \to \mathbb{R}^d$ that satisfies $\langle \phi^\circ(\mathbf{s}), \phi^\circ(\mathbf{t}) \rangle = \mathbb{E}\left[\langle \phi(\mathbf{s}), \phi(\mathbf{t}) \rangle\right]$, where the expectation is over the random variables that determine ϕ. The approximate mapping ϕ can be constructed explicitly by random projections as follows [16]:

$$\phi(\mathbf{t}) = \sqrt{\frac{2}{d}} \left[\cos(v_1^{\mathrm{T}}\mathbf{t} + \omega_1), \ldots, \cos(v_d^{\mathrm{T}}\mathbf{t} + \omega_d)\right]^{\mathrm{T}} \tag{10}$$

where $v_1, \ldots, v_d \in \mathbb{R}^n$ are i.i.d. samples from a distribution with density $p(v)$, and $\omega_1, \ldots, \omega_d \in \mathbb{R}$ are from the uniform distribution on $[0, 2\pi]$. The density function $p(v)$ is determined by the types of the kernels we want to use. For the Gaussian kernel $k^\circ(\mathbf{s}, \mathbf{t}) = \exp(-\sigma\|\mathbf{s}-\mathbf{t}\|_2^2)$, we have $p(v) = (4\pi\sigma)^{-d/2} \exp\left(-\|v\|_2^2/(4\sigma)\right)$, from the Fourier transformation of k°.

This approximation method is less expensive than the previous one, requiring only $O(nd)$ operations for each data point (assuming sampling of each vector $v_i \in \mathbb{R}^n$ takes $O(n)$ time). As we see in Sect. 4, however, this approach tends to give lower prediction accuracy than the first approach for a fixed d value.

2.4 *Efficient Prediction*

Given the solution (γ, b) of (PL), we now describe how the prediction of a new data point $\mathbf{t} \in \mathbb{R}^n$ can be made efficiently without recovering the support vector coefficient α in (P2). The imposed low dimensionality of the approximate kernel in our approach can lead to significantly lower cost of prediction, as low as a fraction of $d/$(no. support vectors) of the cost of an exact-kernel approach.

For the feature mapping approximation, we can simply use the decision function f suggested immediately by (P1), that is, $f(\mathbf{t}) = \mathbf{w}^\mathrm{T}\phi(\mathbf{t}) + b$. Using the definitions (2), (6), and $\gamma := V^\mathrm{T}\alpha$, we obtain

$$f(\mathbf{t}) = \phi(\mathbf{t})^\mathrm{T} \sum_{i=1}^{m} \alpha_i \phi(\mathbf{t}_i) + b = \phi(\mathbf{t})^\mathrm{T} V^\mathrm{T} \alpha + b = \phi(\mathbf{t})^\mathrm{T} \gamma + b.$$

The time complexity in this case is $O(nd)$.

For the kernel matrix approximation approach, the decision function $\mathbf{w}^\mathrm{T}\phi(\mathbf{t}) + b$ cannot be used directly, as we have no way to evaluate $\phi(\mathbf{t})$ for an arbitrary point $\mathbf{t}$. We can, however, use the approximation (5) to note that

$$\phi(\mathbf{t})^\mathrm{T}\mathbf{w} + b = \sum_{i=1}^{m} \alpha_i \phi(\mathbf{t})^\mathrm{T} \phi(\mathbf{t}_i) + b \approx \sum_{i=1}^{m} \alpha_i k^\circ(\mathbf{t}_i, \mathbf{t}) + b, \tag{11}$$

so we can define the function on the right-hand side of this expression to be the decision function. To evaluate this expression, we need to recover a vector α such that $V^\mathrm{T}\alpha = \gamma$, where γ is obtained by solving (PL). Since V^T has dimensions $n \times \bar{d}$, α is not uniquely defined by this expression. Thus, we choose to set $\alpha_i = 0$ for $i \notin \mathcal{S}$, and define the remaining subvector $\alpha_\mathcal{S}$ as follows (in the notation of Sect. 2.3):

$$\alpha_\mathcal{S} := Q_{\cdot,1..\bar{d}} D^{-1/2}_{1..\bar{d},1..\bar{d}} \gamma.$$

By using (9) and the decomposition $(\Psi^\circ)_{\mathcal{S}\mathcal{S}} = QDQ^\mathrm{T}$, it is easy to verify that $V^\mathrm{T}\alpha = V^\mathrm{T}_{\mathcal{S}\cdot}\alpha_\mathcal{S} = \gamma$, as required. Calculation of this α can be performed in $O(d^2)$ time. Therefore, prediction of a test point will take $O(d^2 + nd)$, including kernel evaluation time.

3 The ASSET Algorithm

Here we describe our stochastic approximation approach, with reference to the general convex optimization problem

$$\min_{x \in X} f(x),$$

where f is a convex function and $X \subset \mathbb{R}^d$ is a compact convex set with radius $D_X := \max_{x \in X} ||x||_2$. We assume that at any $x \in X$, we can calculate a stochastic subgradient estimate $G(x;\xi)$ depending on random variable $\xi \in \Xi \subset \mathbb{R}^p$, for which $\mathbb{E}[G(x;\xi)] \in \partial f(x)$. The norm deviation of the stochastic subgradients is measured by D_G defined as follows:

$$\mathbb{E}[||G(x;\xi)||_2^2] \leq D_G^2, \quad \forall x \in X, \ \xi \in \Xi.$$

Iterate Update: At iteration j, the algorithm takes the following step:

$$x^j = \Pi_X(x^{j-1} - \eta_j G(x^{j-1};\xi^j)), \ \ j = 1,2,\ldots,$$

where ξ^j is a random variable (i.i.d. with the random variables used at previous iterations), Π_X is the Euclidean projection onto X, and $\eta_j > 0$ is a step length. For our problem (PL), we have $x^j = (\gamma^j, b^j)$, ξ^j is selected to be one of the indices $\{1,2,\ldots,m\}$ with equal probability, and the subgradient estimate is constructed from the subgradient for the ξ^jth term in the summation of the empirical loss term. Table 1 summarizes the subgradients $G(x^{j-1};\xi^j)$ for classification and regression tasks, with the hinge loss and the ϵ-insensitive loss functions, respectively.

Feasible Sets: For the classification problem, we define the feasible set X to be the Cartesian product of a ball in the γ component with radius $1/\sqrt{\lambda}$ and an interval for the b component, that is, $b \in [-B,B]$, for some constant B. The radius of the ball is derived using strong duality [19, Theorem 1]; we have $D_X = \sqrt{1/\lambda + B^2}$. For the regression problem, the following theorem provides a bound on the size of γ^*, and thus suggests a radius for X.

Table 1 Loss functions and their subgradients for classification and regression

Task	ℓ	$G\left(\begin{bmatrix}\gamma^{j-1}\\ b^{j-1}\end{bmatrix};\xi^j\right) = \begin{bmatrix}\lambda\gamma^{j-1} + d_j V_{\xi^j.}^{\mathrm{T}}\\ d_j\end{bmatrix}$
Classification	$\max\{1 - y(\mathbf{w}^{\mathrm{T}}\phi(\mathbf{t}) + b), 0\}$	$d_j = \begin{cases} -y_{\xi^j} & \text{if } y_{\xi^j}(V_{\xi^j.}\gamma^{j-1} + b^{j-1}) < 1 \\ 0 & \text{otherwise.} \end{cases}$
Regression	$\max\{\lvert y - (\mathbf{w}^{\mathrm{T}}\phi(\mathbf{t}) + b)\rvert - \epsilon, 0\}$	$d_j = \begin{cases} -1 & \text{if } y_{\xi^j} > V_{\xi^j.}\gamma^{j-1} + b^{j-1} + \epsilon, \\ 1 & \text{if } y_{\xi^j} < V_{\xi^j.}\gamma^{j-1} + b^{j-1} - \epsilon, \\ 0 & \text{otherwise.} \end{cases}$

Theorem 1. *For SVM regression using the ϵ-insensitive loss function with $0 \le \epsilon < \|\mathbf{y}\|_\infty$, where $\mathbf{y} := (y_1, y_2, \ldots, y_m)^{\mathrm{T}}$, we have for the optimal γ^* that $\|\gamma^*\|_2 \le \sqrt{2(\|\mathbf{y}\|_\infty - \epsilon)/\lambda}$.*

Proof. We can write an equivalent formulation of (PL) for regression as

$$\min_{\gamma,b} \; \frac{1}{2}\gamma^{\mathrm{T}}\gamma + C\sum_{i=1}^{m} \max\{|y_i - (\gamma^{\mathrm{T}}\phi(\mathbf{t}_i) + b)| - \epsilon, 0\},$$

for $C = 1/(\lambda m)$. The corresponding Lagrange dual formulation is

$$\max_{z,z'} \; -\frac{1}{2}\sum_{i=1}^{m}\sum_{j=1}^{m}(z_i' - z_i)(z_j' - z_j)\langle\phi(\mathbf{t}_i),\phi(\mathbf{t}_j)\rangle - \epsilon\sum_{i=1}^{m}(z_i' + z_i) + \sum_{i=1}^{m} y_i(z_i' - z_i)$$

$$\text{s.t.} \; \sum_{i=1}^{m}(z_i' - z_i) = 0, \; 0 \le z_i \le C, \; 0 \le z_i' \le C, \; i = 1, 2, \ldots, m.$$

Let (γ^*, b^*) and (z^*, z'^*) be the optimal solutions of the primal and the dual formulations, respectively. Replacing $\gamma^* = \sum_{i=1}^{m}(z_i'^* - z_i^*)\phi(\mathbf{t}_i)$ from the KKT conditions into the dual objective and using strong duality, we have

$$\begin{aligned}
\frac{1}{2}(\gamma^*)^{\mathrm{T}}\gamma^* &\le \frac{1}{2}(\gamma^*)^{\mathrm{T}}\gamma^* + C\sum\nolimits_{i=1}^{m} \max\{|y_i - ((\gamma^*)^{\mathrm{T}}\phi(\mathbf{t}_i) + b^*)| - \epsilon, 0\} \\
&= -\frac{1}{2}(\gamma^*)^{\mathrm{T}}\gamma^* - \epsilon\sum\nolimits_{i=1}^{m}(z_i'^* + z_i^*) + \sum\nolimits_{i=1}^{m} y_i(z_i'^* - z_i^*) \\
&\le -\frac{1}{2}(\gamma^*)^{\mathrm{T}}\gamma^* + (\|\mathbf{y}\|_\infty - \epsilon)(\|z\|_1 + \|z'\|_1).
\end{aligned}$$

From the constraints on z, we have $\|z\|_\infty \le C$, and thus $\|z\|_1 \le Cm = 1/\lambda$. (Similarly for z'.) Our claim follows from using these bounds in the inequality above. Note if $\epsilon \ge \|\mathbf{y}\|_\infty$, then the optimal solution is trivially $(\gamma^*, b^*) = (\mathbf{0}, 0)$. □

Averaged Iterates: The solution of (3) is estimated not by the iterates x^j but rather by a weighted sum of the final few iterates. Specifically, if we define N to be the total number of iterates to be used and $\bar{N} < N$ to be the point at which we start averaging, the final reported solution estimate would be $\tilde{x}^{\bar{N},N} := \frac{\sum_{t=\bar{N}}^{N}\eta_t x^t}{\sum_{t=\bar{N}}^{N}\eta_t}$. There is no need to store all the iterates x^t, $t = \bar{N}, \bar{N}+1, \ldots, N$ in order to evaluate the average. Instead, a running average can be maintained over the last $N - \bar{N}$ iterations, requiring the storage of only a single extra vector.

Estimation of D_G: The steplength η_j requires knowledge of the subgradient estimate deviation D_G defined in (3). We use a small random sample of training data indexed by $\xi^{(l)}$, $l = 1, 2, \ldots, M$, at the first iterate (γ^0, b^0), and estimate D_G^2 as $\mathbb{E}[\|G([\gamma^0, b^0]^{\mathrm{T}}; \xi)\|_2^2] \approx \frac{1}{M}\sum_{l=1}^{M} d_l^2(\|V_{\xi^{(l)}\cdot}\|_2^2 + 1)$.

Algorithm 1: ASSET Algorithm

Input: $T = \{(\mathbf{t}_i, y_i)\}_{i=1}^m$, Ψ°, λ, integers $\bar{N}$ and N with $0 < \bar{N} < N$, D_X, D_G.
Set $(\gamma^0, b^0) = (\mathbf{0}, 0)$, $(\tilde{\gamma}, \tilde{b}) = (\mathbf{0}, 0)$, $\tilde{\eta} = 0$;
for $j = 1, 2, \ldots, N$ **do**
  $\eta_j = \frac{D_X}{D_G\sqrt{j}}$;
  Choose $\xi^j \in \{1, \ldots, m\}$ at random;
  $$V_{\xi^j \cdot} = \begin{cases} V_{\xi^j \cdot} & \text{for } V \text{ as in (9), or} \\ \phi(\mathbf{t}_{\xi^j}) & \text{for } \phi(\cdot) \text{ as in (10).} \end{cases}$$
  Compute $G\left(\begin{bmatrix}\gamma^{j-1} \\ b^{j-1}\end{bmatrix}; \xi^j\right)$ following Table 1;
  $$\begin{bmatrix}\gamma^j \\ b^j\end{bmatrix} = \Pi_X\left(\begin{bmatrix}\gamma^{j-1} \\ b^{j-1}\end{bmatrix} - \eta_j G\left(\begin{bmatrix}\gamma^{j-1} \\ b^{j-1}\end{bmatrix}; \xi^j\right)\right).$$
  if $j \geq \bar{N}$ **then** (Update averaged iterate)
  $$\begin{bmatrix}\tilde{\gamma} \\ \tilde{b}\end{bmatrix} = \frac{\tilde{\eta}}{\tilde{\eta} + \eta_j}\begin{bmatrix}\tilde{\gamma} \\ \tilde{b}\end{bmatrix} + \frac{\eta_j}{\tilde{\eta} + \eta_j}\begin{bmatrix}\gamma^j \\ b^j\end{bmatrix}, \quad \tilde{\eta} = \tilde{\eta} + \eta_j.$$
  end
end
Output: $\tilde{\gamma}^{\bar{N},N} := \tilde{\gamma}$ and $\tilde{b}^{\bar{N},N} := \tilde{b}$.

We summarize this framework in Algorithm 1 and refer it as ASSET. The integer $\bar{N} > 0$ specifies the iterate at which the algorithm starts averaging the iterates, which can be set to 1 to average all, to a predetermined maximum iteration number to output just the last one, or to a number in between.

3.1 Convergence

The analysis of robust stochastic approximation [14,15] provides theoretical support for the algorithm above. Considering Algorithm 1 applied to the general formulation (3), choosing $\bar{N} = \lceil \rho N \rceil$ for some specified constant $\rho \in (0,1)$, and denoting the algorithm's output by $\tilde{x}^{\bar{N},N}$, we have the following result.

Theorem 2. *Given the output $\tilde{x}^{\bar{N},N}$ and optimal function value $f(x^*)$, Algorithm 1 satisfies*

$$\mathbb{E}[f(\tilde{x}^{\bar{N},N}) - f(x^*)] \leq C(\rho)\frac{D_X D_G}{\sqrt{N}},$$

where $C(\rho)$ solely depends on the fraction $\rho \in (0,1)$ for which $\bar{N} = \lceil \rho N \rceil$.

3.2 Strongly Convex Case

Suppose that we omit the intercept b from the linear formulation (PL). Then its objective function $f(x)$ becomes strongly convex for all of its variables. In this special case we can apply different steplength $\eta_j = 1/(\lambda j)$ to achieve faster theoretical convergence rate—a rate of $1/j$ rather than $1/\sqrt{j}$. The algorithm remains the same as Algorithm 1 except that averaging is no longer needed. (See [15] for a general proof.)

Theorem 3. *Given the output x^N and optimal function value $f(x^*)$, Algorithm 1 with $\eta_j = 1/(\lambda j)$ satisfies*

$$\mathbb{E}[f(x^N) - f(x^*)] \leq \frac{1}{N} \max\left\{ \left(\frac{D_G}{\lambda}\right)^2, D_X^2 \right\}.$$

Note that when $\lambda \approx 0$ (that is, when the modulus of convexity is small), the convergence of this approach can be quite slow unless we have $D_G \approx 0$ as well.

Without the intercept b, the feasible set X is simpler (as it contains only the γ component), so the update steps are changed accordingly. The resulting algorithm, which we refer to as ASSET*, is the same as Pegasos [18] and SGD [3], apart for our extensions to nonlinear kernels.

4 Computational Results

We implemented Algorithm 1 by modifying the open-source Pegasos code. (Our code is available at http://pages.cs.wisc.edu/~sklee/asset/.) The versions of our algorithms that use kernel matrix approximation are referred to as ASSET$_M$ and ASSET$_M^*$, while those with feature mapping approximation are called ASSET$_F$ and ASSET$_F^*$. For direct comparisons with other codes, we do not include intercept terms since some of the other codes do not allow such terms to be used without penalization. All experiments with randomness are repeated 50 times.

Table 2 summarizes the six binary classification tasks we use, indicating the values of parameters λ and σ selected using SVM-Light to maximize the classification accuracy on each validation set. (For `MNIST-E`, we use the same parameters as in `MNIST`.) For the first five moderate-size tasks, we compare all of our algorithms against four publicly available codes: CPNY and CPSP (both cutting-plane methods), along with SVM-Light and LASVM. (The original SVM-Perf [9] and OCAS [7] are not included because they cannot handle nonlinear kernels.) For `MNIST-E`, we compare our algorithms using feature mapping approximation to LASVM.

For our algorithms, the averaging parameter is set to $\bar{N} = m - 100$ for all cases (averaging is performed for the final 100 iterates). The test error values are computed using the efficient schemes of Sect. 2.4.

Table 2 Data sets and training parameters

Name	m(train)	Valid/test	n	(density)	λ	σ	Note
ADULT	32561	8140/8141	123	(11.2%)	3.07e-8	0.001	UCI Repository
MNIST	58100	5950/5950	784	(19.1%)	1.72e-7	0.01	Digits 0-4 vs. 5-9
CCAT	78127	11575/11574	47237	(1.6%)	1.28e-6	1.0	RCV1 collection [13]
IJCNN	113352	14170/14169	22	(56.5%)	8.82e-8	1.0	2001 challenge[a]
COVTYPE	464809	58102/58101	54	(21.7%)	7.17e-7	1.0	Type 1 vs. rest
MNIST-E	1000000	20000/20000	784	(25.6%)	1.00e-8	0.01	An extended set[b]

[a] http://www.csie.ntu.edu.tw/~cjlin/libsvmtools/datasets/
[b] http://leon.bottou.org/papers/loosli-canu-bottou-2006/

4.1 Accuracy vs. Approximation Dimension

The first experiment investigates the effect of kernel approximation dimension on classification accuracy. We set the dimension parameter s in Sect. 2.3 to values in the range $[2, 1024]$, with the eigenvalue threshold $\epsilon_d = 10^{-16}$. Note that s is an upper bound on the actual dimension d of approximation for ASSET$_M^{(*)}$, but is equal to d in the case of ASSET$_F^{(*)}$. The CPSP and CPNY have a parameter similar to s, as an upper bound of d; we set this parameter to the same values as our s.

For the first five moderate-size tasks, we ran our algorithms for 1,000 epochs ($1000m$ iterations) so that they converged to a near-optimal value with small variation among different randomization. We obtained the baseline performance of these tasks by running SVM-Light. SVM-Light does not have dimension parameters but can be expected to give the best achievable performance by the kernel-approximate algorithms as s approaches m.

Figure 1 shows the results. We do not plot results for ASSET$_M^*$ or ASSET$_F^*$, as they give very similar results to ASSET$_M$ and ASSET$_F$, respectively. When the value of σ is very small, as in Fig. 1(a) of the ADULT data set, all codes achieve good classification performance for small dimension. In other data sets, the chosen values of σ are larger and the intrinsic rank of the kernel matrix is higher, so classification performance continues to improve as s increases.

Interestingly, ASSET$_F$ (feature mapping approximation) seems to require a higher dimension than ASSET$_M$ (kernel matrix approximation) to produce similar classification accuracy. We can, however, afford to use a larger dimension for ASSET$_F$, since the former requires less computation than the latter. For fixed dimension, the overall performance of ASSET$_F$ is worse than other methods, especially in the CCAT experiment.

The cutting plane method CPSP generally requires lower dimension than the others to achieve the same prediction performance. This is because CPSP spends extra time to construct optimal basis functions, whereas the other methods depend on random sampling. However, all approximate-kernel methods including CPSP suffer considerably from the restriction in dimension for the COVTYPE task.

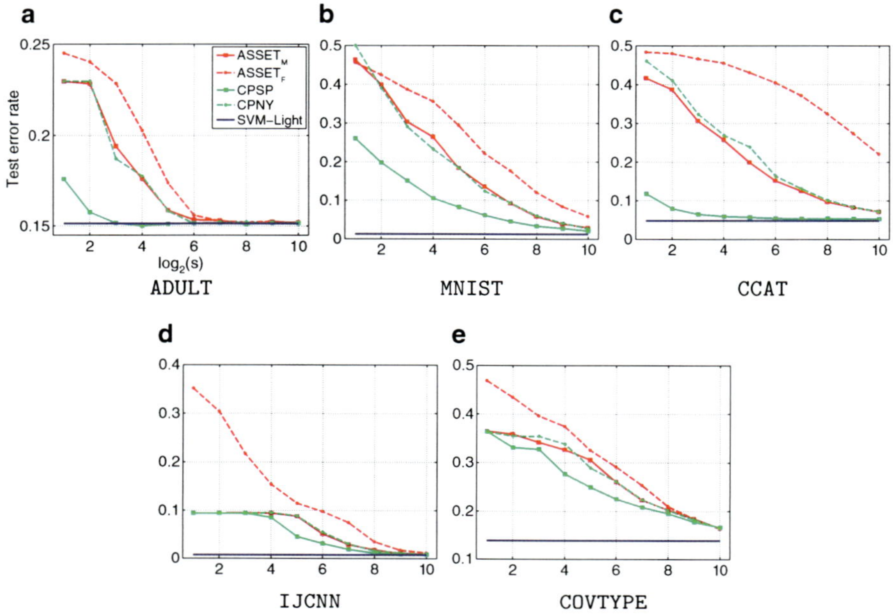

Fig. 1 The effect of the approximation dimension to the test error. The x-axis shows the values of s in log scale (base 2)

4.2 Time Requires to Achieve Similar Test Error

Here we ran all algorithms other than ours with their default stopping criteria. For ASSET_M and ASSET^*_M, we checked the classification error on the test sets ten times per epoch, terminating when the error matched the performance of CPNY. (Since this code uses a similar Nyström approximation of the kernel, it is the one most directly comparable with ours in terms of classification accuracy.) The test error was measured using the iterate averaged over the 100 iterations immediately preceding each checkpoint.

Results for the first five data sets are shown in Table 3 for $s = 512$ and $s = 1,024$. (LASVM and SVM-Light do not depend on s and so their results are the same in both tables.) Our methods are the fastest in most cases. Although the best classification errors among the approximate codes are obtained by CPSP, the runtimes of CPSP are considerably longer than for our methods. In fact, if we compare the performance of ASSET_M with $s = 1,024$ and CPSP with $s = 512$, ASSET_M achieves similar test accuracy to CPSP (except for CCAT) but is faster by a factor between two and forty. CPNY requires an abnormally long run time on ADULT; we surmise that the code may be affected by numerical difficulties.

It is noteworthy that ASSET_M shows similar performance to ASSET^*_M despite the less impressive theoretical convergence rate of the former. This is because the

Table 3 Training CPU time (in seconds, h:hours) and test error rate (%) in parentheses

	Subgradient methods				Cutting-plane				Decomposition			
$s = 512$	$ASSET_M$		$ASSET^*_M$		CPSP		CPNY		LASVM		SVM-Light	
ADULT	23	(15.1±.06)	24	(15.1±.06)	3,020	(15.2)	8.2h	(15.1)	1,011	(18.0)	857	(15.1)
MNIST	97	(4.0±.05)	101	(4.0±.04)	550	(2.7)	348	(4.1)	588	(1.4)	1323	(1.2)
CCAT	95	(8.2±.08)	99	(8.3±.06)	800	(5.2)	62	(8.3)	2,616	(4.7)	3423	(4.7)
IJCNN	87	(1.1±.02)	89	(1.1±.02)	727	(0.8)	320	(1.1)	288	(0.8)	1331	(0.7)
COVTYPE	697	(18.2±.06)	586	(18.2±.07)	1.8h	(17.7)	1,842	(18.2)	38.3h	(13.5)	52.7h	(13.8)
$s = 1024$	$ASSET_M$		$ASSET^*_M$		CPSP		CPNY		LASVM		SVM-Light	
ADULT	78	(15.1±.05)	83	(15.1±.04)	3,399	(15.2)	7.5h	(15.2)	1,011	(18.0)	857	(15.1)
MNIST	275	(2.7±.03)	275	(2.7±.02)	1,273	(2.0)	515	(2.7)	588	(1.4)	1323	(1.2)
CCAT	265	(7.1±.05)	278	(7.1±.04)	2,950	(5.2)	123	(7.2)	2,616	(4.7)	3423	(4.7)
IJCNN	307	(0.8±.02)	297	(0.8±.01)	1,649	(0.8)	598	(0.8)	288	(0.8)	1,331	(0.7)
COVTYPE	2,259	(16.5±.04)	2064	(16.5±.06)	4.1h	(16.6)	3,598	(16.5)	38.3h	(13.5)	52.7h	(13.8)

Kernel approximation dimension is varied by setting $s = 512$ and $s = 1,024$ for $ASSET_M$, $ASSET^*_M$, CPSP, and CPNY. Decomposition methods do not depend on s, so their results are the same in both tables

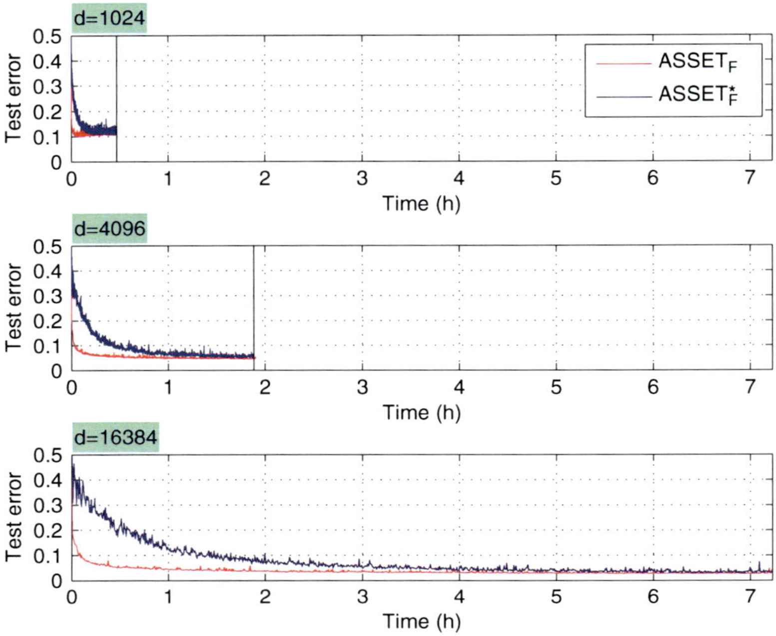

Fig. 2 Progress of ASSET_F and ASSET_F^* in terms of test error rate (MNIST-E)

values of optimal regularization parameter λ were near zero in our experiments, and thus the objective function lost the strong convexity condition required for ASSET_M^* to work. We observed similar slowdown of Pegasos and SGD when λ approaches zero for linear SVMs.

4.3 Large-Scale Performance

We take the final data set MNIST-E and compare the performance of ASSET_F and ASSET_F^* to the online SVM code LASVM. (Other algorithms such as CPSP, CPNY, and SVM-Light are less suitable for large-scale comparison because they operate in batch mode.) For a fair comparison, we fed the training samples to the algorithms in the same order.

Figure 2 shows the progress on a single run of our algorithms, with various approximation dimensions d (which we set equal to s in these experiments) in the range $[1024, 16384]$. Vertical bars in the graphs indicate the completion of training. ASSET_F tends to converge faster and shows smaller test error values than ASSET_F^*, despite the theoretical slower convergence rate of the former. With $d = 16384$, ASSET_F and ASSET_F^* required 7.2 h to finish with a solution of 2.7% and 3.5%

test error rate, respectively. LASVM produced a better solution with only 0.2 % test error rate, but it required 4.3 days of computation to complete a single pass through the same training data.

5 Conclusion

We have proposed a stochastic gradient framework for training large-scale and online SVMs using efficient approximations to nonlinear kernels. Since our approach does not require strong convexity of the objective function or dual reformulations for kernelization, it can be extended easily to other kernel-based learning problems.

Acknowledgements The authors acknowledge the support of NSF Grants DMS-0914524 and DMS-0906818, and in part by Deutsche Forschungsgemeinschaft (DFG) within the Collaborative Research Center SFB 876 "Providing Information by Resource-Constrained Analysis," project C1.

References

1. Bordes, A., Ertekin, S., Weston, J., Bottou, L.: Fast kernel classifiers with online and active learning. J. Mach. Learn. Res. **6**, 1579–1619 (2005)
2. Boser, B.E., Guyon, I.M., Vapnik, V.N.: A training algorithm for optimal margin classifiers. In: Proceedings of the Fifth Annual Workshop on Computational Learning Theory, pp. 144–152. ACM, New York (1992)
3. Bottou, L.: SGD: Stochastic gradient descent, http://leon.bottou.org/projects/sgd (2005). Accessed 30 Mar 2012
4. Chapelle, O.: Training a support vector machine in the primal. Neural Comput. **19**, 1155–1178 (2007)
5. Drineas, P., Mahoney, M.W.: On the Nystrom method for approximating a gram matrix for improved kernel-based learning. J. Mach. Learn. Res. **6**, 2153–2175 (2005)
6. Fan, R.-E., Chen, P.-H., Lin, C.-J.: Working set selection using second order information for training SVM. J. Mach. Learn. Res. **6**, 1889–1918 (2005)
7. Franc, V., Sonnenburg, S.: Optimized cutting plane algorithm for support vector machines. In: Proceedings of the 25th International Conference on Machine Learning, pp. 320–327. ACM, New York (2008)
8. Joachims, T.: Making large-scale support vector machine learning practical. In: Advances in Kernel Methods – Support Vector Learning, pp. 169–184. MIT, Cambridge (1999)
9. Joachims, T.: Training linear SVMs in linear time. In: International Conference on Knowledge Discovery and Data Mining, pp. 217–226. ACM, New York (2006)
10. Joachims, T., Yu, C.-N.J.: Sparse kernel SVMs via cutting-plane training. Mach. Learn. **76**(2–3), 179–193 (2009)
11. Joachims, T., Finley, T., Yu, C.-N.: Cutting-plane training of structural SVMs. Mach. Learn. **77**(1), 27–59 (2009)
12. Kimeldorf, G., Wahba, G.: A correspondence between bayesian estimation on stochastic processes and smoothing by splines. Ann. Math. Statist. **41**, 495–502 (1970)
13. Lewis, D.D., Yang, Y., Rose, T.G., Dietterich, G., Li, F., Li, F.: RCV1: A new benchmark collection for text categorization research. J. Mach. Learn. Res. **5**, 361–397 (2004)

14. Nemirovski, A., Yudin, D.B.: Problem Complexity and Method Efficiency in Optimization. Wiley, New York (1983)
15. Nemirovski, A., Juditsky, A., Lan, G., Shapiro, A.: Robust stochastic approximation approach to stochastic programming. SIAM J. Optim. **19**(4), 1574–1609 (2009)
16. Rahimi, A., Recht, B.: Random features for large-scale kernel machines. In: Advances in Neural Information Processing Systems, vol. 20, pp. 1177–1184. MIT, Cambridge (2008)
17. Scholkopf, B., Smola, A.J.: Learning with Kernels: Support Vector Machines, Regularization, Optimization, and Beyond. MIT, Cambridge (2001)
18. Shalev-Shwartz, S., Singer, Y., Srebro, N.: Pegasos: Primal estimated sub-GrAdient SOlver for SVM. In: Proceedings of the 24th International Conference on Machine Learning, pp. 807–814. ACM, New York (2007)
19. Shalev-Shwartz, S., Singer, Y., Srebro, N., Cotter, A.: Pegasos: Primal estimated sub-GrAdient SOlver for SVM. Math. Program. Ser. B **127**(1), 3–30 (2011)
20. Zinkevich, M.: Online convex programming and generalized infinitesimal gradient ascent. In: Proceedings of the 20th International Conference on Machine Learning, pp. 928–936. ACM, New York (2003)

Single-Frame Signal Recovery Using a Similarity-Prior

Sakinah A. Pitchay and Ata Kabán

Abstract We consider the problem of signal reconstruction from noisy observations in a highly under-determined problem setting. Most of previous work does not consider any specific extra information to recover the signal. Here we address this problem by exploiting the similarity between the signal of interest and a consecutive motionless frame. We incorporate this additional information of similarity that is available into a probabilistic image-prior based on the Pearson type VII Markov Random Field model. Results on both synthetic and real data of MRI images demonstrate the effectiveness of our method in both compressed setting and classical super-resolution experiments.

Keywords Single-frame super-resolution • Compressive sensing • Similarity prior • Image recovery

1 Introduction

Conventional image super-resolution (SR) aims to recover a high-resolution scene from a single or multiple frames of low-resolution measurements. A noisy frame of a single low-resolution image or signal often suffers from a blur and down-sampling transformation. The problem is more challenging when the observed data is a single low-resolution frame that has fewer measurements than the number of unknown pixels in the high-resolution scene that we aim to recover. This makes the problem ill-posed and under-determined too. For this reason, some additional prior knowledge is vital to obtain a satisfactory solution. We have demonstrated in earlier work [10] that the Pearson type VII density integrated with Markov Random Fields (MRF) is an appropriate approach for this purpose.

S. A. Pitchay (✉) • A. Kabán
School of Computer Science, University of Birmingham, Edgbaston, Birminghm B15 2TT, UK
e-mail: S.A.Pitchay@cs.bham.ac.uk, sakinah.ali@usim.edu.my; A.Kaban@cs.bham.ac.uk

P. Latorre Carmona et al. (eds.), *Mathematical Methodologies in Pattern Recognition and Machine Learning*, Springer Proceedings in Mathematics & Statistics 30,
DOI 10.1007/978-1-4614-5076-4_6,

In this paper, we tackle the problem using more specific prior information, namely the similarity to a motionless consecutive frame as the additional input for recovering the signals of interest in a highly under-determined setting. This has real applications, e.g. in medical imaging where such frames are obtained from several scans. Previous work in [13] found the average frame from those scans to be useful for recovery.

In principle, the more information we have about the recovered signal, the better the recovery algorithm is expected to perform. This hypothesis seems to work in [5, 13]; however, both of these works require us to tune the free parameters of the model manually, and [5] reckons that the range of parameter values was not exhaustively tested. Reference [13] also mentions that they were not able to attain exact reconstruction using fewer measurements than those needed by compressed sensing (CS) for a small image. By contrary, in this paper we will demonstrate good recovery from very few measurements using a probabilistic model that includes an automated estimation of its hyper-parameters.

Related work on sparse reconstruction gained tremendous interest recently and can be found in [2–4, 9]. The sparser a signal is, in some basis, the fewer random measurements are sufficient for its recovery. However these works do not consider any specific extra information that could be used to accentuate the sparsity, which is our focus. Somewhat related, the recent work in [11] exploits partial erroneous information to recover small image sequences.

This paper is aimed at taking these ideas further through a more principled and more comprehensive treatment. We consider the case when the observed frame contains too few measurements, but an additional motionless consecutive scene in high resolutions is provided as an extra input. This assumption is often realistic in imaging applications. Our aim is to reduce the requirements on the number of measurements by exploiting the additional similarity information. To achieve this, we employ a probabilistic framework, which allows us to estimate all parameters of our model in an automated manner. We conduct extensive experiments that show that our approach not only bypasses the requirement of tuning free parameters but it is also superior to a cross validation method in terms of both accuracy and computation time.

2 Image Recovery Framework

2.1 Observation Model

A model is good if it explains the data. The following linear model has been used widely to express the degradation process from the high-resolution signal $\mathbf{z}$ to a compressed or low-resolution noisy signal $\mathbf{y}$ [6–8, 12]:

$$\mathbf{y} = \mathbf{W}\mathbf{z} + \eta \tag{1}$$

where the high-resolution signal denoted by $\mathbf{z}$ is an N-dimensional column vector and $\mathbf{y}$ is an Mx1 matrix representing the noisy version of the signal, with $M < N$.

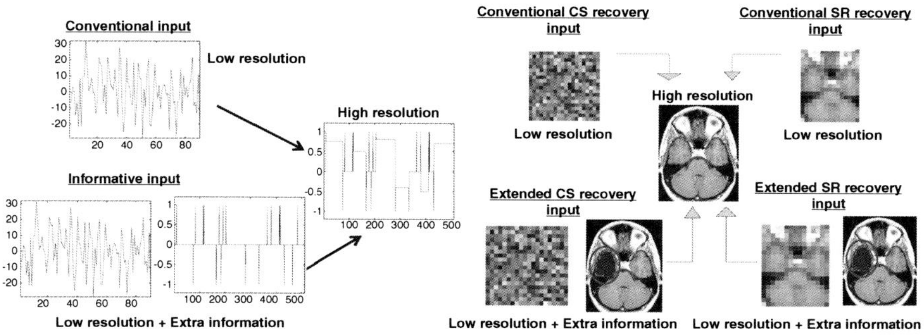

Fig. 1 An illustration of a signal recovery process from a noisy version of low resolution for 1D (*left*) and 2D (*right*) signals with the aid of informative input

In classical super-resolution, the transformation matrix $\mathbf{W}$ typically consists of blur and down-sampling operators. In our study, we also utilise random Gaussian compressive matrices $\mathbf{W}$ with entries sampled independent and identically distributed (i.i.d) from a standard Gaussian. Finally, $\boldsymbol{\eta}$ is the additive noise, assumed to be Gaussian with zero-mean and variance, σ^2.

Before we proceed with the details of the similarity-prior, an example for 1D and 2D signals in Fig. 1 depicts the input and the output for both signals recovery.

2.2 The Similarity-Prior

The construction of a generic prior for images, the Pearson type VII MRF prior was presented in [10]. It is based on the neighbourhood features $\mathbf{Dz}$ where $\mathbf{D}$ makes the signal sparse. In this paper, we aim to recover both 1D and 2D signals using the additional similarity information. We define the entries of $\mathbf{D}$, i.e. d_{ij} as follows:

$$d_{ij} = \begin{cases} 1 & \text{if } i = j; \\ -1/\# & \text{if} i \text{ and} j \text{ are neighbours;} \\ 0 & \text{otherwise.} \end{cases}$$

where # denotes the number of cardinal neighbours and it is 4 for images and 2 for 1D signals.

In general, the idea is that the main characteristic of any natural image is a local-smoothness. This means that the intensities of neighbouring pixels tend to be very similar. Hence, $\mathbf{Dz}$ will be sparse. Therefore, here we propose an enhanced prior to exploit more information that leads to more sparseness. By employing the given additional information of the consecutive image or signal, we will employ the difference, $\mathbf{f}$ between the recovered image, $\mathbf{z}$ and the extra information denoted as $\mathbf{s}$. Obviously the more pixels $\mathbf{z}$ and $\mathbf{s}$ have in common, the more smooth their difference will be. Figure 2 shows a few examples of histograms of the neighbourhood features

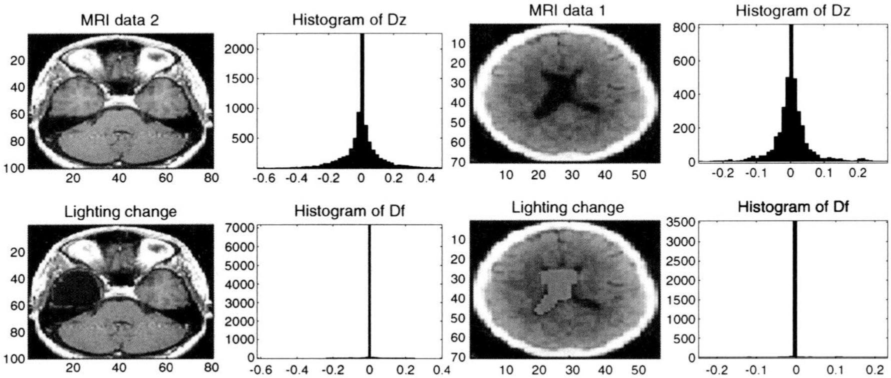

Fig. 2 Example histograms of the distribution of neighbourhood features $\mathbf{D}_i\mathbf{z}$, and $\mathbf{D}_i\mathbf{f}$ where $i = 1,\ldots,N$ from an MRI real data

Dz from real images, where the sparsity is entirely the consequence of the local smoothness. Additionally, we also show the histograms of the new neighbourhood features **Df** that includes the additional similarity information. We see the latter is a lot sparser than the former.

Then we can formulate the ith feature in a vector form, with the aid of the ith row of this matrix (denoted $\mathbf{D}_i$) as the following:

$$f_i - \frac{1}{\#} \sum_{j \in \# \text{ neighb}(i)} f_j = \sum_{j=1}^{N} d_{ij} f_j = \mathbf{D}_i \mathbf{f} \tag{2}$$

Since our task is to encode the sparse property of signals, this feature is useful: The difference between a pixel of the difference image f and the average of its neighbours is close to zero, almost everywhere except at the edges of the dissimilarity areas. Plugging this into the Pearson-MRF density, we have the following prior that we refer to as a *similarity-prior*:

$$\Pr(\mathbf{z}) = \frac{1}{Z_{\Pr(\lambda,v)}} \prod_{i=1}^{N} \{(\mathbf{D}_i(\mathbf{z}-\mathbf{s}))^2 + \lambda\}^{-\frac{1+v}{2}} \tag{3}$$

where $Z_{\Pr(\lambda,v)} = \int \mathrm{d}z \prod_{i=1}^{N} \{(\mathbf{D}_i(\mathbf{z}-\mathbf{s}))^2 + \lambda\}^{-\frac{1+v}{2}}$ is the partition function that makes the whole probability density function integrate to one, and this multivariate integral does not have an analytic form.

2.3 *Pseudo-likelihood Approximation*

As in previous work [10], we employ a pseudo-likelihood approximation to the partition function $Z_{p(\lambda,v)}$. Replacing the approximation using the extra information into (3), we obtain the following approximate image model:

$$\Pr(\mathbf{z}|\lambda,\nu) \approx \prod_{i=1}^{N} \frac{\Gamma\left(\frac{1+\nu}{2}\right)\lambda^{\nu/2}\{(\mathbf{D}_i(\mathbf{z}-\mathbf{s}))^2+\lambda\}^{-\frac{1+\nu}{2}}}{\Gamma\left(\frac{\nu}{2}\right)\sqrt{\pi}} \tag{4}$$

We shall employ this to infer $\mathbf{z}$ simultaneously with estimating our hyper-parameters λ, ν and σ.

2.4 *Joint Model*

The entire model is the joint model of the observations $\mathbf{y}$ and the unknowns $\mathbf{z}$.

$$\Pr(\mathbf{y},\mathbf{z},f|\mathbf{W},\sigma^2,\lambda,\nu) = \Pr(\mathbf{y}|\mathbf{z},\mathbf{W},\sigma^2)\Pr(\mathbf{z}|f,\lambda,\nu) \tag{5}$$

where the first factor is the observation model and the second factor is the image prior model and its free parameters defined as λ and ν.

3 MAP Estimation

We will employ the joint probability (5) as the objective to be maximised. Maximising this w.r.t. $\mathbf{z}$ is also equivalent to finding the most probable image $\hat{\mathbf{z}}$, i.e. the maximum a posteriori (MAP) estimate, since (5) is proportional to the posterior $\Pr(\mathbf{z}|\mathbf{y})$.

$$\hat{\mathbf{z}} = \arg\min_{\mathbf{z}}\{-\log[\Pr(\mathbf{y}|\mathbf{z})] - \log[\Pr(\mathbf{z})]\} \tag{6}$$

Namely, the most probable high-resolution signal is the one for which the negative log of the joint probability model takes its minimum value. Hence, our problem can be solved through minimisation. The expression for the negative log of the joint probability model will then be defined as our minimisation objective and also called as the error-objective. It can be written as:

$$\mathrm{Obj}(\mathbf{z},\sigma^2,\lambda,\nu) = -\log[\Pr(\mathbf{y}|\mathbf{z},\sigma^2)] - \log[\Pr(\mathbf{z}|f,\lambda,\nu)] \tag{7}$$

Equation (7) may be decomposed into two terms: the first one that contains all the entries that involve $\mathbf{z}$ and the second one contains the terms that do not—i.e. $\mathrm{Obj}(\mathbf{z},\sigma^2,\lambda,\nu) = \mathrm{Obj}_z(\mathbf{z}) + \mathrm{Obj}_{(\lambda,\nu)}(\lambda,\nu)$.

3.1 *Estimating the Most Probable z*

The observation model is also called the likelihood model because it expresses how likely it is that a given $\mathbf{z}$ produced the observed $\mathbf{y}$ through the transformation $\mathbf{W}$. Hence we have for the first term in (5):

$$\Pr(\mathbf{y}|\mathbf{z}) \propto \exp\left\{-\frac{1}{2\sigma^2}(\mathbf{y}-\mathbf{Wz})^{\mathrm{T}}(\mathbf{y}-\mathbf{Wz})\right\} \tag{8}$$

By plugging in the term for the observation model and the prior into (7), we obtain the objective function. The terms of the objective (7) that depend on $\mathbf{z}$ are the following:

$$\mathrm{Obj}_z(\mathbf{z}) = \frac{1}{2\sigma^2}(\mathbf{y}-\mathbf{Wz})^2 + \frac{\nu+1}{2}\sum_{i=1}^{N}\log\left\{(D_i(\mathbf{z}-\mathbf{s}))^2+\lambda\right\} \tag{9}$$

The most probable estimate is the $\hat{z}$ that has the highest probability in the model. It is equivalently the one that achieves the lowest error. Recap, our model has two factors which depend on the likelihood or also known as the observation model, and the image prior that assists the signal recovery. Thus, our error models both the *mismatch* of the predicted model $\mathbf{Wz}$ with the observed data $\mathbf{y}$ and *determinant* for allowing the free parameters to control the smoothness and the edges encoded in the prior. The objective is differentiable; therefore, any non-linear optimiser could be practical to optimise the term (9) w.r.t. $\mathbf{z}$. The gradient of the negative log likelihood term is given by:

$$\nabla(z)\mathrm{Obj}_z = \frac{1}{\sigma^2}\mathbf{W}'(\mathbf{Wz}-\mathbf{y}) + (\nu+1)\sum_{i=1}^{N} D_i^{\mathrm{T}}\frac{D_i(\mathbf{z}-\mathbf{s})}{(D_i(\mathbf{z}-\mathbf{s}))^2+\lambda} \tag{10}$$

3.2 Estimation of σ^2, λ and ν

Writing out the terms in (7) that depend on σ^2, we obtain a closed form for estimating the σ^2.

$$\sigma^2 = \frac{1}{M}\left(\sum_{i=1}^{M}(y_i - \mathbf{W}_i\mathbf{z})^2\right) \tag{11}$$

Terms that depend on λ and ν are given by:

$$\begin{aligned}\mathrm{Obj}_{(\lambda,\nu)} =& N\log\Gamma\left(\frac{1+\nu}{2}\right) - N\log\Gamma\left(\frac{\nu}{2}\right) + \frac{N\nu}{2}\log\lambda \\ & -\frac{1+\nu}{2}\sum_{i=1}^{N}\log((\mathbf{D}_i(\mathbf{z}-\mathbf{s}))^2+\lambda)\end{aligned} \tag{12}$$

Both of these hyperparameters need to be positive valued. To ensure our estimates are actually positive, we parameterise the log probability objective (12) such as to optimise for the $+/-$ square root of these parameters. Taking derivatives w.r.t $\sqrt{\lambda}$ and $\sqrt{\nu}$, we obtain:

$$\frac{\mathrm{d}\log p(\mathbf{z})}{\mathrm{d}\sqrt{\lambda}} = \sum_{i=1}^{N} \frac{\nu(\mathbf{D}_i(\mathbf{z}-\mathbf{s}))^2 - \lambda}{((\mathbf{D}_i(\mathbf{z}-\mathbf{s}))^2 + \lambda)\sqrt{\lambda}} \tag{13}$$

$$\frac{\mathrm{d}\log p(\mathbf{z})}{\mathrm{d}\sqrt{\nu}} = \left[N\log\lambda - \sum_{i=1}^{N}\log((\mathbf{D}_i(\mathbf{z}-\mathbf{s}))^2 + \lambda) + N\psi\left(\frac{1+\nu}{2}\right) - N\psi\frac{\nu}{2}\right]\sqrt{\nu} \tag{14}$$

where $\psi(.)$ is the digamma function. The zeros of these functions give us the estimates of $\pm\sqrt{\lambda}$ and $\pm\sqrt{\nu}$. Although there is no closed-form solution, these can be obtained numerically using any unconstrained non-linear optimisation method,[1] which requires the gradient vector of the objectives.

3.3 Recovery Algorithm

Our algorithm implements the equations given in the previous section as described in Algorithm 1. At each iteration of the algorithm, two smaller gradient descent problems have to be solved; namely one for λ, ν and one for $\mathbf{z}$. However, experiment suggests that it is not necessary to estimate the minimum with high accuracy. We notice that the inner loops do not require the entire convergence. It is sufficient to increase but not necessarily minimise the objective at each intermediate step.

Algorithm 1 : Recovery algorithm

```
1: Initialise the estimates z
2: iterate until convergence: do
3:     estimate σ² using (11)
4:     iteratively update λ and ν in turn using definiton
5:    (13) and (14), with the current estimate z.
6:     iterate to update z using (10)
7: end
```

[1] We made use of the efficient implementation available from http://www.kyb.tuebingen.mpg.de/bs/people/carl/code/minimize/.

4 Experiments and Discussion

We devise two hypotheses as following to investigate the role of the new prior and we test those using synthetic 1D and 2D signals and real MRI signals:

1. The quality of the recovered signal using the additional information is no worse than the one without the extra information provided that the extra information is *useful*. This is when the number of zero entries in the new form of the neighbourhood feature, i.e. **Df** is larger than the number of zero entries in **Dz**, that is the generic feature that has not been given the extra similarity information.
2. The fewer the edges in **f** (that is, the non-zeros in **Df**), the fewer measurements are sufficient for enabling a successful recovery.

We should mention the construction of the measurement matrix **W** from CS-type W is a random Gaussian matrix ($M \times N$) with *iid* entries. The SR-type W is a deterministic transformation that blurs and down-samples the image.[2]

4.1 *Illustrative 1D Experiments*

In this section, we implement our recovery algorithm on the 1D data, derived from a spike signal[3] of size 512×1 as shown in Fig. 3a. We proceed by plugging the extra signal into our image prior and varying the number of measurements using randomly generated measurement matrices **W** with *iid* Gaussian entries as in CS. The recovery results are summarised in Fig. 3b. We see our enhanced prior is capable to achieve a good recovery and has a lower mean square error (MSE) than the one without extra information.

We also examine the MSE performance as a function of the number of zero entries in the relevant feature vectors (i.e. **Df** in our case). Figure 4 shows MSE results when varying the number of zero entries by constructing variations on the signals. We see when the recovery algorithm received sufficient measurements, for example when $M = 250$ in Fig. 3, the role of the proposed *similarity prior* gradually reduces. In other words, this *similarity prior* is *useful* in massively under-determined problems and provided that the given extra information has the characteristics described previously.

A widely used alternative way to set hyperparameters is cross-validation. It is therefore of interest how does the automated estimation of the hyper-parameters of our Pearson type VII-based MRF compare to a cross-validation procedure. Next, we address this by looking at two aspects: MSE performance and CPU time. We use the

---

[2]Code to generate the SR-type matrices can be found from http://www.robots.ox.ac.uk/~elle/SRcode/index.html.

[3]Data is taken from http://people.ee.duke.edu/~lcarin/BCS.html.

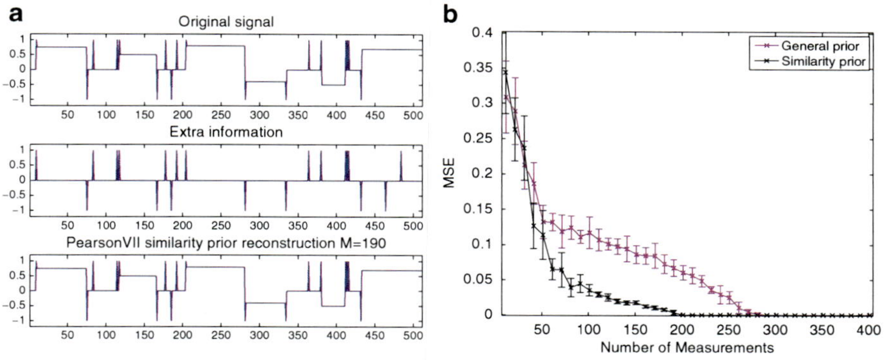

Fig. 3 (**a**) The original spike signal; the extra similarity information; and an example of recovered signal from 190 measurements. (**b**) Comparing the MSE performance of 1D spike signal recovery with and without the extra information. The error bars are over ten independent trials and the level of noise was $\sigma = 8e - 5$

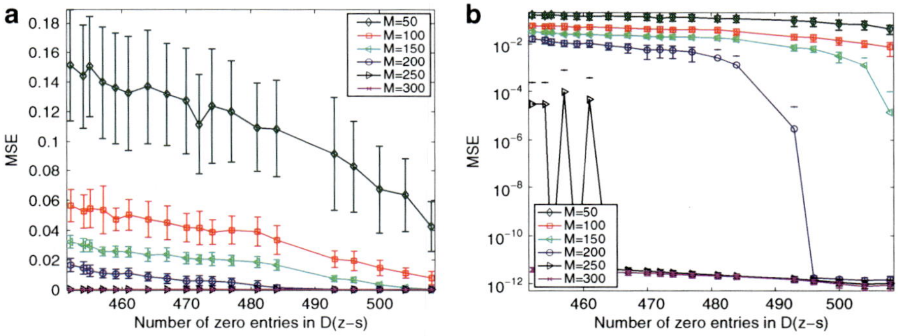

Fig. 4 (**a**) Linear scale. (**b**) Log scale. MSE performance of 1D spike signal using the extra information. The number of zero entries in **D**(**z**-**s**) is varied. The error bars represent one standard error about the mean from 50 independent trials. The level of noise was $\sigma = 8e - 5$

same spike signal for this purpose. For our comparison, we have chosen fivefolds cross validation method for estimating the hyper-parameters λ and ν and the noise variance is assumed to be known for this method. A sensible search range is pursued to avoid a long execution time as we are aware that this method can be extremely time-consuming if the search space is too large.

Figure 5 shows the MSE performance and the associated values for the four levels of noise using the CS-type **W**. It is interesting to see that our fully automated parameter estimation turns out to be superior to fivefolds cross validation and it has fast convergence and less execution time.

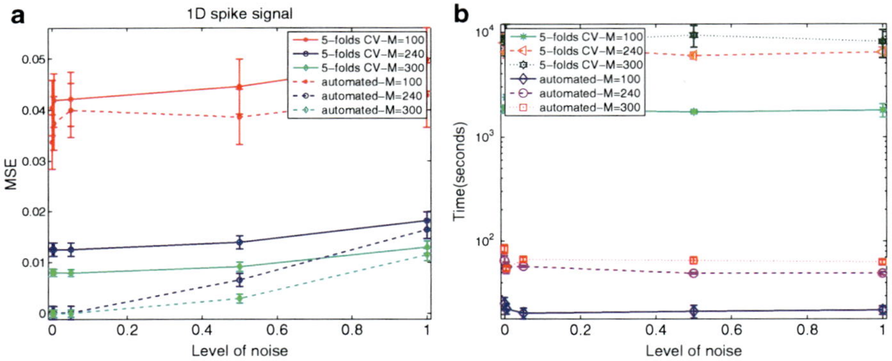

Fig. 5 (**a**) Comparing the MSE performance of the fully automated Pearson type VII-based MRF approach with the fivefolds cross validation, tested with four levels of noise (σ= 0.005, 0.05, 0.5, 1). (**b**) Cpu time performance against the same four levels of noise. We see that our automated estimation and recovery is significantly faster than the fivefolds cross validation method. The error bars are over ten repeated trials for each level of noise. Three sets of measurements ($M = 100, 240, 300$) have been tested for this accuracy comparison

4.2 2D Experiments

Following the thorough understanding gained in the previous section about when the extra information is helpful on the spike signal test cases, we conducted experiments with both compressive sensing (CS) matrices where W contains random entries and also the classical super-resolution matrices where W consists of blur and down-sampling. In this set of experiments, we consider a motionless scene as the extra information. More precisely, the extra information that we employ in our similarity-prior consists of a change in the lighting of some area in the image.

We start by conducting the recovery algorithm on a synthetic data of size $[50 \times 50]$. The noise variance σ tested in all experiments is set to a smaller range in order to tally the general noise in real data.

Figures 6 and 7 show examples of vastly under-determined problems using the extra information for recovery in comparison with the previous prior devised in [10]. The MSE performance results are given in Fig. 8, and we see the MSE drops rapidly with increasing the measurement size. Figure 9 shows examples of recovered images from this process. We observe that the quality of the recovered image increases rapidly for all five levels of noise tested. This is in contrast with the recovery results from the general prior, which needs a lot more measurements to perform well.

From these findings, the degree of similarity of the available extra information has a significant impact on the recovery from insufficient measurements. We find that without informative extra information the recovery algorithm does not perform well with such few measurements. The recovered signal and the MSE using the artificial *Phantom* data in Figs. 6 and 8 demonstrate that the fewer the

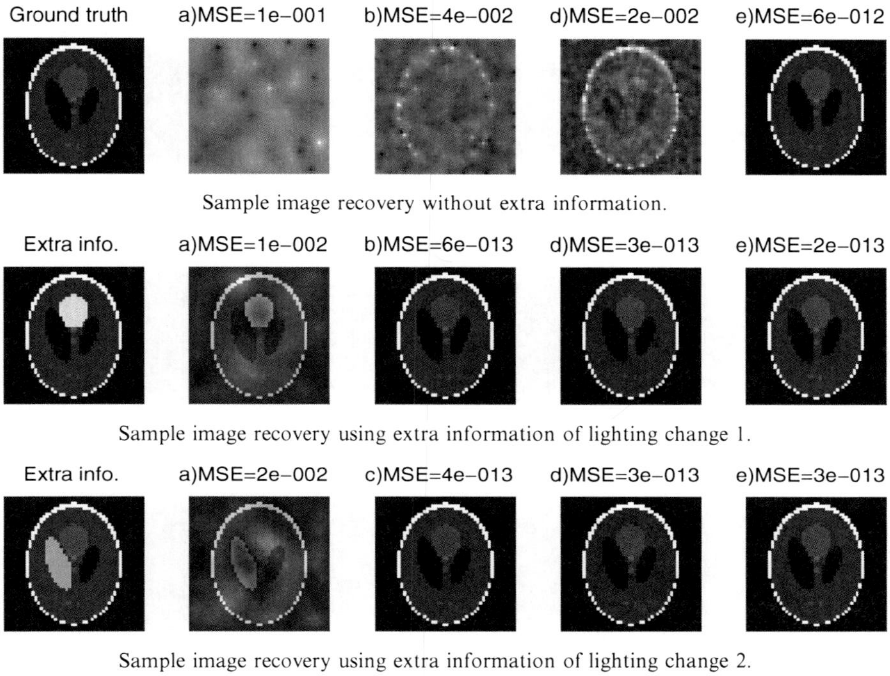

Fig. 6 Example recovery of 2D synthetic data of size [50 × 50] in the case of using SR-type W, and given two slightly different light changes as extra similarity information. The number of measurements (M) are: (**a**) $M = 60$, (**b**) 460, (**c**) 510, (**d**) 960, (**e**) 1,310. The additive noise level was $\sigma = 8e - 5$

edges in the difference image **f** the better the recovery, or the smaller the number of measurements needed for a good recovery. This result validates our second hypothesis.

In the remainder of the experiments, we will now focus on image recovery using real image data of magnetic resonance imaging (MRI). We obtained this data from the Matlab database and we created the additional similarity information from it by changing the lighting of an area on the image. Next we validate our second hypothesis on a variety of MRI images and its lighting changes. The recovery results for both types of W are presented in Figs. 11 and 12. The MSE performance for the CS-type W is shown in Fig. 10. Interestingly, we observe that the log scale in that figure is in more direct correspondence with our visual perception rather than using the standard linear scale, and this will be seen by comparison with Figs. 11 and 12.

We observed that more than 6,000 measurements are required for a good recovery without the extra information in this example. However, from these results we see that our similarity prior achieves high quality recovery from an order of magnitude less measurements. The recovered images are presented in Figs. 11 and 12 for

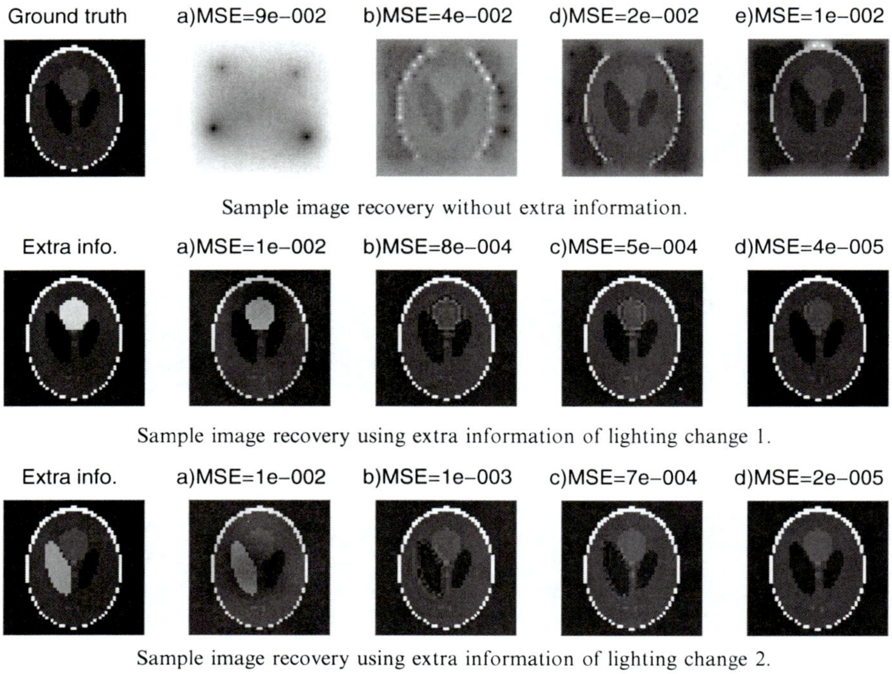

Fig. 7 Example recovery of 2D synthetic data of size [50 × 50] in the case of using SR-type W, and given two slightly different light changes as extra similarity information. The number of measurements (M) are: (**a**) $M = 9$, (**b**) 441, (**c**) 784, (**d**) 1,296, (**e**) 1,849. The additive noise level was $\sigma = 8e - 7$

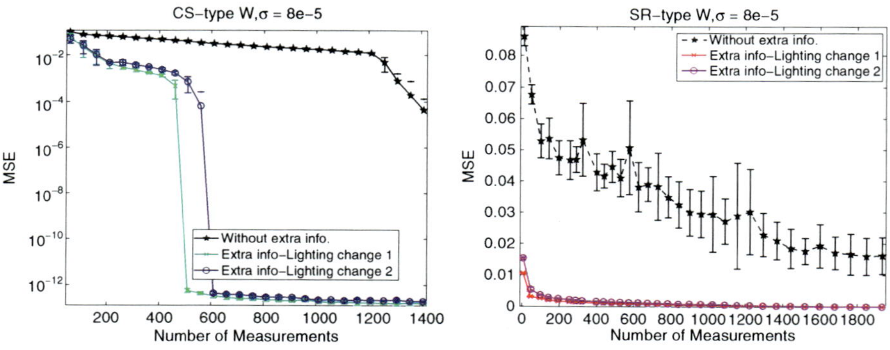

Fig. 8 MSE performance of synthetic data [50 × 50] in comparison with the two types of extra information. Here, both types of W were tested and the noise standard deviation was $\sigma = 8e - 5$

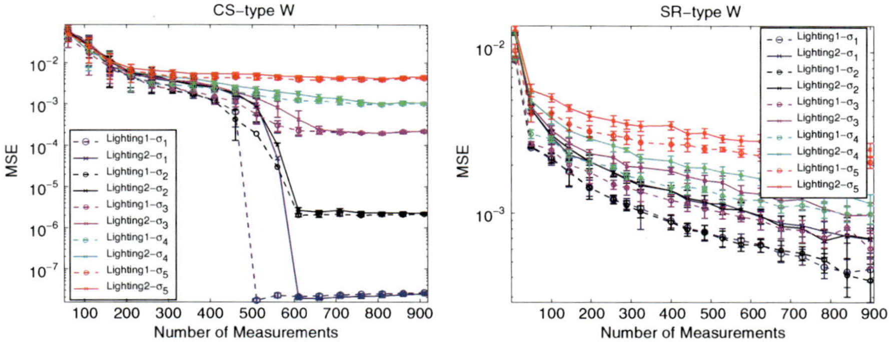

Fig. 9 Recovery of a 50×50 size image from random measurements (*top*) and blurred and down-sampled measurement (*bottom*). The MSE is shown on log scale against varying the number of measurements, in five different levels of noise conditions. The noise levels were as follows. *Top*: $\sigma \in \{\sigma_1 = 0.005,\ \sigma_2 = 0.05,\ \sigma_3 = 0.5,\ \sigma_4 = 1,\ \sigma_5 = 2\}$; *Bottom*: $\{\sigma_1 = 8e - 5,\ \sigma_2 = 8e - 4,\ \sigma_3 = 8e - 3,\ \sigma_4 = 0.016,\ \sigma_4 = 0.032\}$—that is the previous noise levels were divided by $0.8\ \sqrt{N}$ to make the signal-to-noise ratios roughly the same for the two measurement matrix types

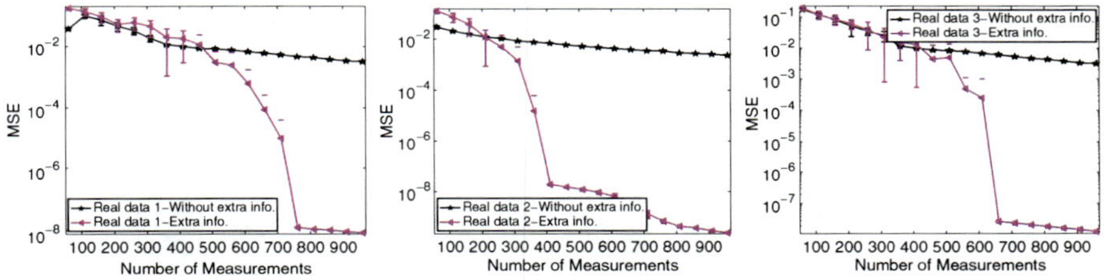

Fig. 10 From *left*: MSE performance of real MRI images of size $[70 \times 57], [70 \times 57]$ and $[100 \times 80]$ in comparison with three types of extra information on the three different sets of data. CS-type W was used and the noise standard deviation was $\sigma = 8e - 5$

a visual comparison. Finally, we also show an example run of our automated parameter estimation algorithm in Fig. 13 for completeness. As one would expect, the speed of convergence varies with the difficulty of the problem.

In closing, we should comment on the possibility of using the other types of extra information for signal recovery. Throughout this paper we exploited the similarity created by a lighting change. Depending on the application domain, one might consider a small shift or rotation instead. However, we have seen that the key for the extra information to be useful in our similarity prior is that the difference image must have fewer edges than the original image. This is not the case with shifts or rotations. Therefore to make such extra information useful we would need to include an image registration model into the prior. This is subject to future work.

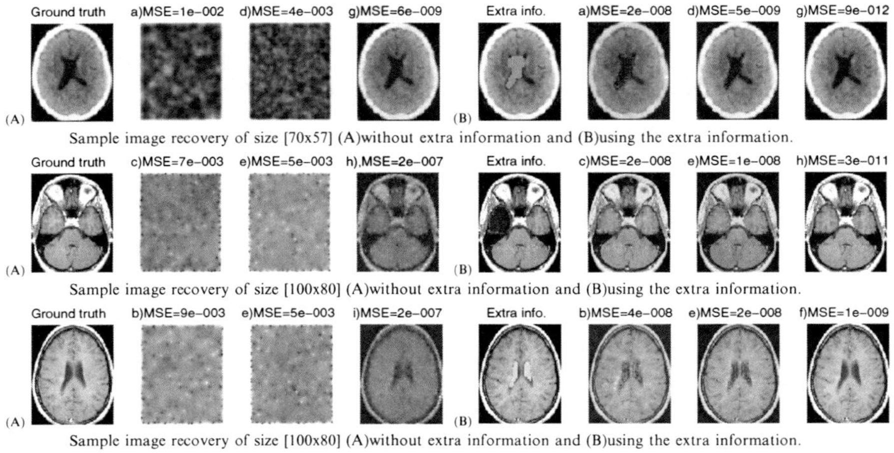

Fig. 11 Examples of MRI image recovery in the case CS-type W, given a motionless consecutive frame with some contrast changes. The number of measurements (M) were: (**a**) $M = 310$, (**b**) 460, (**c**) 560, (**d**) 610, (**e**) 760, (**f**) 1310, (**g**) 3010, (**h**) 5610 (**i**) 7610 and additive noise with $\sigma = 8e-5$. The first row refers to the real data 1, the second row refers to the real data 2 and the third row refers to the real data 3

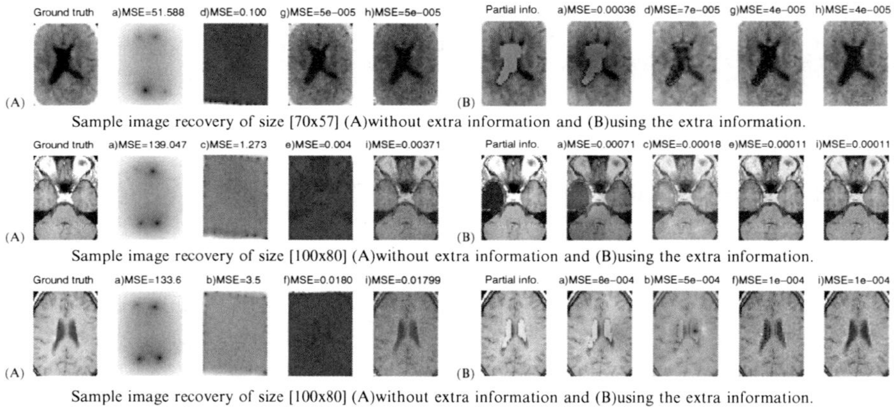

Fig. 12 Examples of MRI image recovery in the case of SR-type W, given a motionless consecutive frame with some contrast changes. The number of measurements (M) were: (**a**) $M = 6$, (**b**) 99, (**c**) 154, (**d**) 396, (**e**) 918, (**f**) 1462, (**g**) 1505, (**h**) 2000, (**i**) 4234. The additive noise is $\sigma = 8e-5$

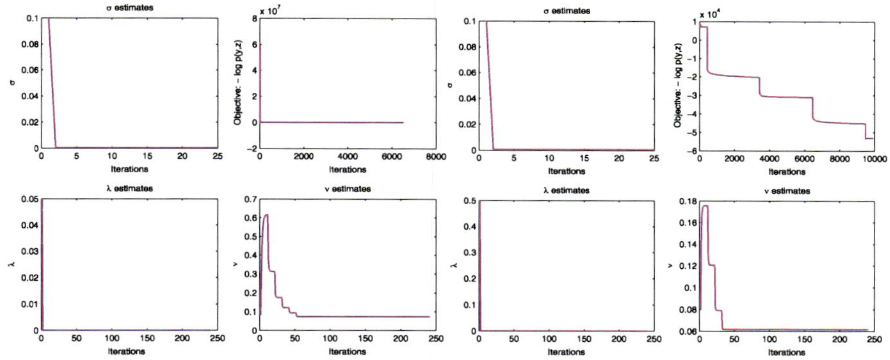

Fig. 13 Example evolution of the hyper-parameter updates (σ, λ, v) and objective function versus the number of iterations of the optimisation algorithm while recovering a 2D signal: from *left*, random measurements; and from *right*, a blurred and down-sampled low-resolution frame. In both experiments, the noise level is $\sigma = 8e - 5$

5 Conclusions

In this paper, we have formulated and employed a similarity-prior-based Pearson type VII Markov Random Field to include the similarity information between the scene of interest and a consecutive scene that has a lighting change. This prior enables us to recover the high-resolution scene of interest from fewer measurements than a general-purpose prior would, and this can be applied, e.g. in medical imaging applications.

Acknowledgements The first author wishes to thank *Universiti Sains Islam Malaysia* (USIM) and the Ministry of Higher Education of Malaysia (MOHE) for the support and facilities provided. Extended version of the work presented at ICPRAM2012 [1].

References

1. Ali Pitchay, S., Kabán, A.: Single-frame signal recovery using a similarity-prior based on Pearson type VII MRF. In: Proceedings of the 1st International Conference on Pattern Recognition Applications and Methods (ICPRAM), pp. 123–133. SciTePres, Vilamoura, Portugal (2012)
2. Baraniuk, R.G., Cevher, V., Duarte, M.F., Hegde, C.: Model-based compressive sensing. In: IEEE Transactions on Information Theory, vol. 56, pp. 1982–2001, Dec 2010, IEEE Press Piscataway, NJ, USA
3. Candes, E., Romberg, J., Tao, T.: Robust uncertainty principles: exact signal reconstruction from highly incomplete frequency information. In: IEEE Transactions on Information Theory, vol. 52, no. 2, pp. 489–509, Feb 2006, IEEE Press Piscataway, NJ, USA
4. Donoho, D.L.: Compressed sensing. In: IEEE Transactions on Information Theory, vol. 52, no. 4, pp. 1289–1306, April 2006, IEEE Press Piscataway, NJ, USA

5. Giraldo, J.C.R., Trzasko, J.D., Leng, S., McCollough, C.H., Manduca, A.: Non-convex prior image constrained compressed sensing (NC-PICCS). In: Proceedings of SPIE: Physics of Medical Imaging, vol. 7622 (2010), Society of Photo-optical Instrumentation Engineers, Bellingham, ETATS-UNIS
6. Hardie, R.C., Barnard, K.J.: Joint MAP registration and high-resolution image estimation using a sequence of undersampled images. IEEE Trans. Image Process. **6**(12), 621–633 (1997).
7. He, H., Kondi, L.P.: MAP based resolution enhancement of video sequences using a Huber-Markov random field image prior model. In: IEEE Conference of Image Processing, pp. 933–936 (2003) IEEE Signal Processing Society
8. He, H., Kondi, L.P.: Choice of threshold of the Huber-Markov prior in MAP based video resolution enhancement. In: IEEE Electrical and Computer Engineering Canadian Conference, vol. 2, pp. 801–804, Nov 2004, IEEE Conference Publications
9. Ji, S., Xue, Y., Carin, L.: Bayesian compressive sensing. In: IEEE Transactions on Signal Processing, vol. 56, no. 6, pp. 2346–2356, June 2008, IEEE Press Piscataway, NJ, USA
10. Kabán, A., Ali Pitchay, S.: Single-frame image recovery using a Pearson type VII MRF. Neurocomputing (Special issue MLSP 2010), Elsevier Science Publisher B. V. Amsterdam, The Netherlands, vol. 80, pp. 111–119, Mar 2012
11. Lu, W., Vaswani, N.: Regularized modified BPDN for noisy sparse reconstruction with partial erroneous support and signal value knowledge. In: IEEE Transactions on Signal Processing, vol. 60, pp. 182–196, Oct 2011, IEEE Press Piscataway, NJ, USA
12. Pickup, L.C., Capel, D.P., Roberts, S.J., Zissermann, A.: Bayesian methods for image super-resolution. Comput. J., vol. 52, no. 1, pp. 101–113, Jan 2009, Oxford University Press, Oxford, UK
13. Vaswani, N., Lu, W.: Modified-CS: modifying compressive sensing for problems with partially known support. In: IEEE Transactions on Signal Processing, vol. 58, no. 9, Sept 2010, IEEE Press Piscataway, NJ, USA

A Discretized Newton Flow for Time-Varying Linear Inverse Problems

Martin Kleinsteuber and Simon Hawe

Abstract The reconstruction of a signal from only a few measurements, deconvolving, or denoising are only a few interesting signal processing applications that can be formulated as linear inverse problems. Commonly, one overcomes the ill-posedness of such problems by finding solutions that match some prior assumptions on the signal best. These are often sparsity assumptions as in the theory of Compressive Sensing. In this paper, we propose a method to track the solutions of linear inverse problems, and consider the two conceptually different approaches based on the synthesis and the analysis signal model. We assume that the corresponding solutions vary smoothly over time. A discretized Newton flow allows to incorporate the time varying information for tracking and predicting the subsequent solution. This prediction requires to solve a linear system of equations, which is in general computationally cheaper than solving a new inverse problem. It may also serve as an additional prior that takes the smooth variation of the solutions into account, or as an initial guess for the preceding reconstruction. We exemplify our approach with the reconstruction of a compressively sampled synthetic video sequence.

1 Introduction

Linear inverse problems arise in various signal processing applications like in signal deconvolution [4], denoising [9], inpainting [3], or signal reconstruction from few indirect measurements as in Compressive Sensing [5, 7]. Basically, the goal is to

M. Kleinsteuber (✉) • S. Hawe
Department of Electrical Engineering and Information Technology, Technische Universität München, Arcisstrasse 21, 80333 München, Germany
e-mail: kleinsteuber@tum.de; simon.hawe@tum.de; http://www.gol.ei.tum.de

P. Latorre Carmona et al. (eds.), *Mathematical Methodologies in Pattern Recognition and Machine Learning*, Springer Proceedings in Mathematics & Statistics 30,
DOI 10.1007/978-1-4614-5076-4_7, © Springer Science+Business Media New York 2013

compute or reconstruct a signal $\mathbf{s} \in \mathbb{R}^n$ from a set of measurements $\mathbf{y} \in \mathbb{R}^m$, with m being less or equal to n. Formally, this measurement process can be written as

$$\mathbf{y} = \mathcal{A}\mathbf{s} + \mathbf{e}, \tag{1}$$

where the vector $\mathbf{e} \in \mathbb{R}^m$ models sampling errors and noise, and $\mathcal{A} \in \mathbb{R}^{m \times n}$ is the measurement matrix. In most interesting cases, recovering $\mathbf{s}$ from the measurements $\mathbf{y}$ is ill-posed because either the exact measurement process and hence $\mathcal{A}$ is unknown as in blind signal deconvolution, or the number of observations is much smaller than the dimension of the signal, which is the case in Compressive Sensing. In this paper, we restrict to the latter case where the measurement matrix $\mathcal{A}$ is known.

To overcome the ill-posedness of this problem and to stabilize the solution, prior assumptions on the signal can be exploited. In this paper, we discuss two conceptually different approaches, the so-called *synthesis* and the *analysis* signal model, cf. [11].

1.1 The Synthesis and the Analysis Model

One assumption that has proven to be successful in signal recovery, cf. [10] is that natural signals admit a sparse representation $\mathbf{x} \in \mathbb{R}^d$ over some dictionary $\mathcal{D} \in \mathbb{R}^{n \times d}$ with $d \geq n$. We say that a vector $\mathbf{x}$ is sparse when most of its coefficients are equal to zero or small in magnitude. When $\mathbf{s}$ admits a sparse representation over $\mathcal{D}$, it can be expressed as a linear combination of only very few atoms $\{\mathbf{d}_i\}_{i=1}^d$, the columns of $\mathcal{D}$, which reads as

$$\mathbf{s} = \mathcal{D}\mathbf{x}. \tag{2}$$

For $d > n$, the dictionary is said to be overcomplete or redundant, consequently the representation $\mathbf{x}$ is not unique.

Now, an approximate solution $\mathbf{s}^\star$ to the original signal can be obtained from the measurements $\mathbf{y}$ by first solving

$$\mathbf{x}^\star = \arg\min_{\mathbf{x} \in \mathbb{R}^d} \quad g(\mathbf{x}) \quad \text{subject to } \|\mathcal{A}\mathcal{D}\mathbf{x} - \mathbf{y}\|_2^2 \leq \epsilon, \tag{3}$$

and afterwards synthesizing the signal from the computed sparse coefficients via $\mathbf{s}^\star = \mathcal{D}\mathbf{x}^\star$. As the signal is synthesized from the sparse coefficients, the reconstruction model (3) is called the *synthesis* reconstruction model [11]. Therein, $g : \mathbb{R}^n \mapsto \mathbb{R}$ is a function that promotes or measures sparsity and $\epsilon \in \mathbb{R}^+$ is an estimated upper bound on the noise power $\|\mathbf{e}\|_2^2$. Although the choice of the ℓ_1-norm for g as a sparseness prior leads to well-behaved convex optimization problems and to perfect signal recovery under certain assumptions, cf. [8], it has been shown in [6] that in most cases, the concave ℓ_p-pseudo-norm

$$\|\mathbf{v}\|_p^p := \sum_i |v_i|^p, \tag{4}$$

with $0 < p < 1$ severely outperforms its convex counterpart. For the presented approach of tracking the solutions of time-varying linear inverse problems, we do not assume convexity of g but we require differentiability. This is why we employ a smooth approximation of the ℓ_p-pseudo-norm. Generally, to find a solution of Problem (3), various algorithms based on convex or non-convex optimization, greedy pursuit methods, or the Bayesian framework exist that use different choices for g. For a broad overview of such algorithms, we refer the interested reader to [20].

Besides utilizing the synthesis model (3) for signal reconstruction, an alternative way to reconstruct $\mathbf{s}$ is given via

$$\begin{aligned}\mathbf{s}^\star = \underset{\mathbf{s}\in\mathbb{R}^n}{\arg\min} \quad & g(\Omega\mathbf{s}) \\ \text{subject to } & \|\mathcal{A}\mathbf{s}-\mathbf{y}\|_2^2 \le \epsilon,\end{aligned} \tag{5}$$

which is known as the *analysis model* [11]. In this model, $\Omega \in \mathbb{R}^{k\times n}$ with $k \geq n$ is called the *analysis operator* and the *analyzed vector* $\Omega\mathbf{s} \in \mathbb{R}^k$ is assumed to be sparse, where sparsity is again measured via an appropriate function g. In contrast to the synthesis model where a signal is fully described by the nonzero elements of $\mathbf{x}$, in the analysis model the zero elements of $\Omega\mathbf{s}$ contain the interesting information. To emphasize this difference between the two models, the term cosparsity has been introduced in [14], which simply counts the number of zero elements of $\Omega\mathbf{s}$. Certainly, as the sparsity in the synthesis model depends on the chosen dictionary, the cosparsity of a signal solely depends on the choice of the analysis operator Ω.

Different analysis operators for image signals proposed in the literature include fused Lasso [19], the translation invariant wavelet transform [18], and probably best known the finite difference operator closely related to the total variation [17].

1.2 Our Contribution

Here, we propose an approach based on minimizing a time-variant version of the unconstrained Lagrangian forms of (3) and (5), which are given by

$$\underset{\mathbf{x}\in\mathbb{R}^d}{\text{minimize}} \quad f_s(\mathbf{x}) = \frac{1}{2}\|\mathcal{A}\mathcal{D}\mathbf{x}-\mathbf{y}\|_2^2 + \lambda g(\mathbf{x}). \tag{6}$$

and

$$\underset{\mathbf{s}\in\mathbb{R}^n}{\text{minimize}} \quad f_a(\mathbf{s}) = \frac{1}{2}\|\mathcal{A}\mathbf{s}-\mathbf{y}\|_2^2 + \lambda g(\Omega\mathbf{s}) \tag{7}$$

respectively. In both formulations, the Lagrange multiplier $\lambda \in \mathbb{R}_0^+$ weighs between the sparsity of the solution and its fidelity to the acquired samples according to the assumed amount of noise in the measurements $\lambda \sim \epsilon$.

Consider now a sequence of linear inverse problems whose solutions vary smoothly over time. As an example, one may think of the denoising short video sequences (without cut), or the reconstruction of compressively sensed magnetic resonance image sequences, cf. [13]. In this work, we propose an approach to track the solutions of such time-varying linear inverse problems. Therefore, we employ preceding solutions to predict the current signal's estimate without acquiring new measurements. To the best of the authors' knowledge, this idea has not been pursued so far in the literature. The crucial idea is to use a discretized Newton flow to track solutions of a time-varying version of (6) and (7). We provide three practical update formulas for the tracking problem and consider both the analysis and the synthesis model. We conclude with an experiment by applying our approach to a short synthetic video sequence.

2 Tracking the Solutions

2.1 Problem Statement

Let $t \mapsto \mathbf{s}(t) \in \mathbb{R}^n$ be a C^1-curve, i.e. having a continuous first derivative which represents a time-varying signal $\mathbf{s}$. Moreover, let $\mathbf{y}(t) = \mathcal{A}\mathbf{s}(t)$ be the measurements of $\mathbf{s}$ at time t. In this paper, we consider the problem of reconstructing a sequence of signals $\left(\mathbf{s}(t_k)\right)_{k\in\mathbb{N}}$ at consecutive instances of time. Instead of estimating $\mathbf{s}(t_{k+1})$ by solving the inverse problem based on the measurements $\mathbf{y}(t_{k+1})$, we investigate in how far the previously recovered estimates $\mathbf{s}_i^\star$ of $\mathbf{s}(t_i)$, $i = 1,\dots,k$ can be employed to *predict* $\mathbf{s}(t_{k+1})$ without acquiring new measurements $\mathbf{y}(t_{k+1})$. This prediction step may serve as an intermediate replacement for this reconstruction step or it may be employed as an initialization for reconstruction at time t_{k+1}. Note that in our approach, we assume a fixed measurement matrix $\mathcal{A}$.

Now, consider the time variant version of the unconstrained Lagrangian functions from (6) and (7), which read asΩ

$$f_s(\mathbf{x},t) = \frac{1}{2}\|\mathcal{A}\mathcal{D}\mathbf{x} - \mathbf{y}(t)\|_2^2 + \lambda g(\mathbf{x}) \tag{8}$$

and

$$f_a(\mathbf{s},t) = \frac{1}{2}\|\mathcal{A}\mathbf{s} - \mathbf{y}(t)\|_2^2 + \lambda g(\Omega\mathbf{s}). \tag{9}$$

For a unified notation, we use $f(\mathbf{z},t)$ to refer to both (8) and (9) simultaneously. Now, for a fixed time t, the gradient

$$F(\mathbf{z},t) := \frac{\partial}{\partial \mathbf{z}} f(\mathbf{z},t) \tag{10}$$

must be zero for an optimal estimate $\mathbf{z}$. Consequently, we want to find the smooth curve $\mathbf{z}(t)$ as illustrated in Fig. 1 such that

$$F(\mathbf{z}(t),t) = 0. \tag{11}$$

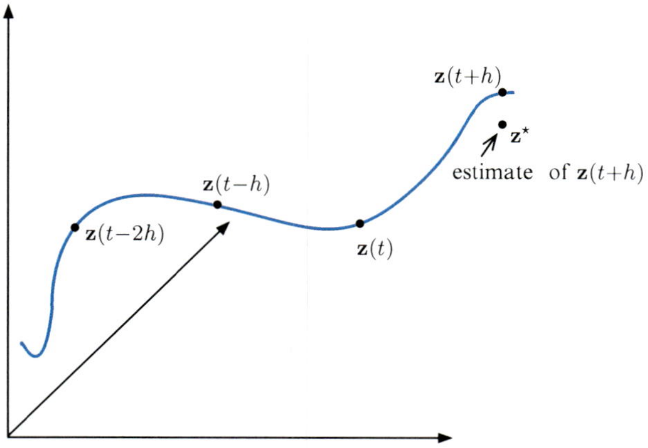

Fig. 1 Smoothly time-varying $\mathbf{z}(t)$. Depending on the applied reconstruction model, $\mathbf{z}(t)$ denotes either the signal itself, or its transform coefficient over the used dictionary

In other words, we want to track the minima of (8) and (9) over time. To achieve this, we employ a discretized Newton flow, which is explained in the following subsection.

2.2 *Discretized Newton Flow*

Homotopy methods are a well-known approach for solving problem (11). These methods are based on an associated differential equation whose solutions track the roots of F. To make the paper self-contained, we shortly rederive the discretized Newton flow for our situation at hand based on [1]. Specifically, we consider the implicit differential equation

$$\mathcal{J}_F(\mathbf{z},t)\dot{\mathbf{z}} + \frac{\partial}{\partial t}F(\mathbf{z},t) = -\alpha F(\mathbf{z},t), \tag{12}$$

where $\alpha > 0$ is a free parameter that stabilizes the dynamics around the desired solution. Here,

$$\mathcal{J}_F(\mathbf{z},t) := \frac{\partial}{\partial \mathbf{z}}F(\mathbf{z},t) \tag{13}$$

is the $(n \times n)$-matrix of partial derivatives of F with respect to $\mathbf{z}$. Under suitable invertibility conditions on $\mathcal{J}_F$, we rewrite (12) in explicit form as

$$\dot{\mathbf{z}} = -\mathcal{J}_F(\mathbf{z},t)^{-1}\left(\alpha F(\mathbf{z},t) + \frac{\partial}{\partial t}F(\mathbf{z},t)\right). \tag{14}$$

We discretize (14) at time instances t_k, for $k \in \mathbb{N}$ and assume without loss of generality a fixed stepsize $h > 0$. Depending on the stepsize we choose $\alpha := \frac{1}{h}$.

With the shorthand notation for $\mathbf{z}_k := \mathbf{z}(t_k)$, the single-step Euler discretization of the time-varying Newton flow is therefore given as

$$\mathbf{z}_{k+1} = \mathbf{z}_k - \mathcal{J}_F(\mathbf{z}_k, t_k)^{-1}\left(F(\mathbf{z}_k, t_k) + h\frac{\partial F}{\partial t}(\mathbf{z}_k, t_k)\right). \tag{15}$$

We approximate the partial derivative $\frac{\partial F}{\partial t}(\mathbf{z}_k, t_k)$ by an mth-order Taylor approximation written as $H_m(\mathbf{z}, t)$. For the practically interesting cases these are

$$H_1(\mathbf{z}, t) = \frac{1}{h}\Big(F(\mathbf{z}, t) - F(\mathbf{z}, t-h)\Big) \tag{16}$$

$$H_2(\mathbf{z}, t) = \frac{1}{2h}\Big(3F(\mathbf{z}, t) - 4F(\mathbf{z}, t-h) + F(\mathbf{z}, t-2h)\Big) \tag{17}$$

$$H_3(\mathbf{z}, t) = \frac{1}{30h}\Big(37F(\mathbf{z}, t) - 45F(\mathbf{z}, t-h) + 9F(\mathbf{z}, t-2h) - F(\mathbf{z}, t-3h)\Big), \tag{18}$$

see also [1]. These approximations turn (15) into the update formula

$$\mathbf{z}^{\star}_{k+1} = \mathbf{z}_k - \mathcal{J}_F(\mathbf{z}_k, t_k)^{-1}\Big(F(\mathbf{z}_k, t_k) + hH_m(\mathbf{z}_k, t_k)\Big). \tag{19}$$

Practically, the inverse $\mathcal{J}_F(\mathbf{z}_k, t_k)^{-1}$ is not accessible or infeasible to calculate, in particular when dealing with high-dimensional data. Hence for computing the estimate $\mathbf{z}^{\star}_{k+1}$ as in (19), we solve

$$\underset{\mathbf{z}\in\mathbb{R}^n}{\text{minimize}} \quad \|\mathcal{J}_F(\mathbf{z}_k, t_k)\mathbf{z} - \mathbf{b}_m(\mathbf{z}_k, t_k)\|_2^2, \tag{20}$$

with

$$\mathbf{b}_m(\mathbf{z}_k, t_k) := \mathcal{J}_F(\mathbf{z}_k, t_k)\mathbf{z}_k - \Big(F(\mathbf{z}_k, t_k) + hH_m(\mathbf{z}_k, t_k)\Big). \tag{21}$$

Typically, linear Conjugate Gradient methods efficiently solve this linear equation, cf. [15]. Note, that this is significantly less computationally expensive than solving an individual reconstruction problem.

In the next subsection, we derive three explicit update schemes for the concrete problem of tracking solutions to inverse problems based on the approximations (16)–(18).

2.3 *Explicit Update Formulas for the Synthesis Model*

Although the previous sections are general enough to deal with any (smooth) sparsity measure g, we want to make our ideas more concrete and employ a concrete smooth approximation of the ℓ_p-pseudo-norm, namely

$$g(\mathbf{x}) = \sum_{i=1}^{d} (x_i^2 + \mu)^{\frac{p}{2}}, \tag{22}$$

with $0 < p \leq 1$ and a smoothing parameter $\mu \in \mathbb{R}^+$. The gradient of g is

$$\nabla g(\mathbf{x}) = p \sum_{i=1}^{d} \mathcal{E}_i \left(x_i^2 + \mu\right)^{\frac{p}{2}-1} \mathbf{x}, \tag{23}$$

where $\mathcal{E}_i := \mathbf{e}_i \mathbf{e}_i^\top$ and $\mathbf{e}_i \in \mathbb{R}^d$ is the standard basis vector. The Hessian of g is given by the diagonal matrix

$$\mathcal{H}_g(\mathbf{x}) = p \sum_{i=1}^{d} \mathcal{E}_i \Big(\left(x_i^2 + \mu\right)^{\frac{p}{2}-1} + (p-2) \left(x_i^2 + \mu\right)^{\frac{p}{2}-2} x_i^2 \Big). \tag{24}$$

Now recall, that in the synthesis model, we have

$$F(\mathbf{x},t) = \frac{\partial}{\partial \mathbf{x}} f_s(\mathbf{x},t) = \mathcal{D}^\top \mathcal{A}^\top (\mathcal{A}\mathcal{D}\mathbf{x} - \mathbf{y}(t)) + \lambda \nabla g(\mathbf{x}). \tag{25}$$

The derivative of F with respect to $\mathbf{x}$ is thus

$$\mathcal{J}_F(\mathbf{x},t) = (\mathcal{A}\mathcal{D})^\top (\mathcal{A}\mathcal{D}) + \lambda \mathcal{H}_g(\mathbf{x}). \tag{26}$$

Analogously as above, for the mth-order Taylor approximation, $m = 1,2,3$, we have

$$hH_1(\mathbf{x},t) = (\mathcal{A}\mathcal{D})^\top \Big(\mathbf{y}(t-h) - \mathbf{y}(t)\Big) \tag{27}$$

$$hH_2(\mathbf{x},t) = \frac{1}{2} (\mathcal{A}\mathcal{D})^\top \Big(4\mathbf{y}(t-h) - 3\mathbf{y}(t) - \mathbf{y}(t-2h)\Big) \tag{28}$$

$$hH_3(\mathbf{x},t) = \frac{1}{30} (\mathcal{A}\mathcal{D})^\top \Big(45\mathbf{y}(t-h) - 37\mathbf{y}(t) - 9\mathbf{y}(t-2h) + \mathbf{y}(t-3h)\Big). \tag{29}$$

This results in the explicit formulas for $\mathbf{b}_1, \mathbf{b}_2, \mathbf{b}_3$

$$\mathbf{b}_1(\mathbf{x}_k,t_k) = \lambda \Big(\mathcal{H}_g(\mathbf{x}_k)\mathbf{x}_k - \nabla g(\mathbf{x}_k)\Big) + (\mathcal{A}\mathcal{D})^\top \Big(2\mathbf{y}(t_k) + \mathbf{y}(t_{k-1})\Big) \tag{30}$$

$$\begin{aligned} \mathbf{b}_2(\mathbf{x}_k,t_k) &= \lambda \Big(\mathcal{H}_g(\mathbf{x}_k)\mathbf{x}_k - \nabla g(\mathbf{x}_k)\Big) \\ &\quad + \frac{1}{2} (\mathcal{A}\mathcal{D})^\top \Big(5\mathbf{y}(t_k) - 4\mathbf{y}(t_{k-1}) + \mathbf{y}(t_{k-2})\Big) \end{aligned} \tag{31}$$

$$\begin{aligned} \mathbf{b}_3(\mathbf{x}_k,t_k) &= \lambda \Big(\mathcal{H}_g(\mathbf{x}_k)\mathbf{x}_k - \nabla g(\mathbf{x}_k)\Big) + \frac{1}{30} (\mathcal{A}\mathcal{D})^\top \Big(67\mathbf{y}(t_k) - 45\mathbf{y}(t_{k-1}) \\ &\quad + 9\mathbf{y}(t_{k-2}) - \mathbf{y}(t_{k-3})\Big). \end{aligned} \tag{32}$$

The three different explicit update formulas for the estimation of the signal at the next instance of time follow straightforwardly as

$$\mathbf{s}_{k+1}^\star = \mathcal{D} \Big\{ \underset{\mathbf{x} \in \mathbb{R}^d}{\arg\min} \, \| \mathcal{J}_F(\mathbf{x}_k,t_k)\mathbf{x} - \mathbf{b}_m(\mathbf{x}_k,t_k) \|_2^2 \Big\}, \quad m = 1,2,3. \tag{33}$$

2.4 Explicit Update Formulas for the Analysis Model

For the analysis model, we use the same sparsity measure g as defined in (22). Let the analysis operator be of dimension $\Omega \in \mathbb{R}^{k\times n}$. We use the notation $(g\circ\Omega)(\mathbf{s}) := g(\Omega\mathbf{s})$ for the composed function. The gradient of $g\circ\Omega$ is

$$\nabla(g\circ\Omega)(\mathbf{s}) = p\Omega^\top \sum_{i=1}^{k} \mathcal{E}_i \left((\mathbf{e}_i^\top \Omega\mathbf{s})^2 + \mu\right)^{\frac{p}{2}-1} \Omega\mathbf{s}. \tag{34}$$

As in the previous section, we have to compute the Hessian of $g\circ\Omega$, which is given by the matrix

$$\mathcal{H}_{(g\circ\Omega)}(\mathbf{s}) = p\Omega^\top \sum_{i=1}^{k} \mathcal{E}_i\Big(\left((\mathbf{e}_i^\top \Omega\mathbf{s})^2 + \mu\right)^{\frac{p}{2}-1} + (p-2)\left((\mathbf{e}_i^\top \Omega\mathbf{s})^2 + \mu\right)^{\frac{p}{2}-2} (\mathbf{e}_i^\top \Omega\mathbf{s})^2\Big)\Omega. \tag{35}$$

Note, that in contrast to the synthesis model, here the Hessian is not diagonal. Equation (10) reads as

$$F(\mathbf{s},t) = \frac{\partial}{\partial \mathbf{s}} f_a(\mathbf{s},t) = \mathcal{A}^\top(\mathcal{A}\mathbf{s} - \mathbf{y}(t)) + \lambda\nabla(g\circ\Omega)(\mathbf{s}) \tag{36}$$

with its derivative with respect to $\mathbf{s}$ being

$$\mathcal{J}_F(\mathbf{s},t) = \mathcal{A}^\top\mathcal{A} + \lambda\mathcal{H}_{(g\circ\Omega)}(\mathbf{x}). \tag{37}$$

Combining (36) with (16)–(18) yields

$$hH_1(\mathbf{s},t) = \mathcal{A}^\top\Big(\mathbf{y}(t-h) - \mathbf{y}(t)\Big) \tag{38}$$

$$hH_2(\mathbf{s},t) = \frac{1}{2}\mathcal{A}^\top\Big(4\mathbf{y}(t-h) - 3\mathbf{y}(t) - \mathbf{y}(t-2h)\Big) \tag{39}$$

$$hH_3(\mathbf{s},t) = \frac{1}{30}\mathcal{A}^\top\Big(45\mathbf{y}(t-h) - 37\mathbf{y}(t) - 9\mathbf{y}(t-2h) + \mathbf{y}(t-3h)\Big). \tag{40}$$

The explicit formulas for $\mathbf{b}_1, \mathbf{b}_2, \mathbf{b}_3$ now result accordingly to the previous subsection. Finally, the explicit update formulas for estimating the signal are

$$\mathbf{s}_{k+1}^\star = \arg\min_{\mathbf{s}\in\mathbb{R}^n} \|\mathcal{J}_F(\mathbf{s}_k,t_k)\mathbf{s} - \mathbf{b}_m(\mathbf{s}_k,t_k)\|_2^2, \quad m = 1,2,3. \tag{41}$$

3 Experiments

In this section we provide an example that should serve as a proof of concept of our proposed algorithm. It consists of tracking the reconstruction result of a series

Fig. 2 Time sequence of synthetic test image

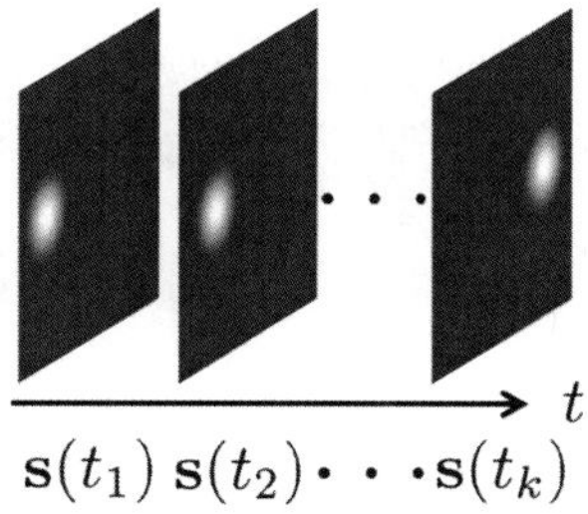

of compressively sampled time-varying images $\mathbf{s}(t_k) \in \mathbb{R}^n$. The images are created synthetically and show a ball moving with constant velocity, see Fig. 2. To enhance legibility, all formulas are expressed in terms of matrix vector products. However, regarding the implementation, we want to emphasize that filtering techniques are used to deal with the large image data.

Considering the measurement matrix $\mathcal{A}$, we chose $m \ll n$ randomly selected coefficients of the Rudin–Shapiro transformation (RST) [2]. The RST, also known as the real-valued Dragon-Noiselet-transformation, is used because of its efficient implementation and due to its desirable properties for image reconstruction [16]. We empirically set the number of measurements to $m = 0.2n$. In our experiments we found that the number of measurements does not severely affect the accuracy of the tracking algorithm, but the speed of convergence. The larger we chose m the faster the algorithm converges.

For the reconstruction, we employ the above discussed analysis model. Therein, the analysis operator $\Omega \in \mathbb{R}^{2n \times n}$ represents a common and simple approximation of the image gradient, which is in terms of finite differences between neighboring pixels in horizontal and vertical directions, respectively.

We start our tracking algorithm by measuring RST coefficients at consecutive instances of time $y(t_k) = \mathcal{A}\mathbf{s}(t_k)$. From these consecutive measurements we find $\mathbf{s}_k^\star$ by individually solving (9) using a nonlinear Conjugate Gradient (CG) method with backtracking line-search and Hestenes-Stiefel update rule, see [12] for the concrete algorithm. From this, we obtain $\mathbf{s}_{k+1}^\star$ by (33), using a linear CG-method. Regarding the update formula for $\mathbf{b}_m$, we found in our experiments that (31) yields a good trade-off between prediction results and computational burden.

The tracking results for our example are presented in Fig. 3b–f for $p = 0.7$. We use the knowledge of $\mathbf{s}(t_k)$, $\mathbf{s}(t_{k-1})$ and $\mathbf{s}(t_{k-2})$ to iteratively estimate $\mathbf{s}_{k+j}^\star$ for $j = 1, \ldots, 5$ only based on the update formula (33). Clearly, the smaller j is, the better the estimation. Note that the results shown in Fig. 3e and f are solely based on previously *predicted* images. The green circle indicates the position of the ball in the original images $\mathbf{s}(t_{k+j})$, $j = 1, \ldots, 5$. It can be seen that although the quality of the images decreases, the position of the circle is still captured adequately. As a quantitative measure of the reconstruction quality, Table 1 contains the peak

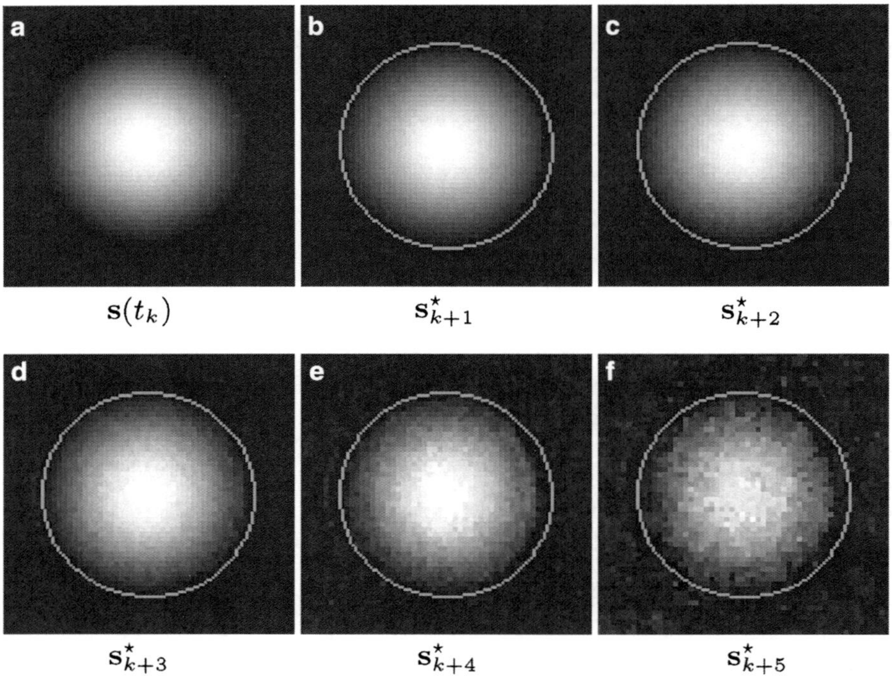

Fig. 3 Excerpt of original image (**a**) and estimated images (**b**)–(**f**). The *green circle* indicates the position of the ball in the original images

Table 1 Peak signal-to-noise ratio (*PSNR*) in decibels (dB) and mean squared error (*MSE*) between estimated signal $\mathbf{s}^\star_{k+j}$ for $j = 1,\dots,5$ and original signals $\mathbf{s}(t_{k+j})$ $j = 1,\dots,5$

	$\mathbf{s}^\star_{k+1}$	$\mathbf{s}^\star_{k+2}$	$\mathbf{s}^\star_{k+3}$	$\mathbf{s}^\star_{k+4}$	$\mathbf{s}^\star_{k+5}$
PSNR	57.2	51.5	34.9	33.3	29.0
MSE	0.12	0.45	20.8	30.3	80.2

signal-to-noise ratio (*PSNR*) $\mathrm{PSNR} = 10\log\left(\frac{\max(\mathbf{s})^2 n}{\sum_{i=1}^{n}(s_i - s_i^\star)}\right)$ and the mean squared error (MSE) $MSE = \frac{1}{n}\sum_{i=1}^{n}(s_i - s_i^\star)^2$ of the estimated signals $\mathbf{s}^\star$ to the original signals $\mathbf{s}$.

A final word on the computational cost of the algorithm. Within the analysis reconstruction model, the cost for applying the Hessian operator as defined in (24) mainly depends on the cost of applying Ω and its transpose, since the remaining part is just a diagonal operator that can be applied in $O(n)$ flops.

Furthermore, we want to mention that for both signal reconstruction models the presented algorithm does not depend on a specific sparsifying transformation $\mathcal{D}$, or analysis operator Ω, respectively. Any transformation or operator that admits a fast implementation, e.g. the Wavelet or Curvelet transformation, or the finite difference operator for images, can be easily used within this framework.

4 Conclusion

In this paper we present a concept for tracking the solutions of inverse problems that vary smoothly over time. We consider the two related but conceptually different synthesis and analysis signal reconstruction models. The tracking is achieved by employing a discretized Newton flow on the gradient of the cost function. The approach allows us to predict the signal at the next time instance from previous reconstruction results without explicitly taking new measurements. One advantage is that this prediction step is computationally less expensive than an individual reconstruction. Furthermore, it may be employed as an initialization, or serve as an additional prior for solving an inverse problem at time t_k.

Acknowledgments This work has partially been supported by the Cluster of Excellence *CoTeSys* – Cognition for Technical Systems, funded by the German Research Foundation (DFG).

References

1. Baumann, M., Helmke, U., Manton, J.: Reliable tracking algorithms for principal and minor eigenvector computations. In: 44th IEEE Conference on Decision and Control and European Control Conference, Institute of Electrical and Electronics Engineers pp. 7258–7263 (2005)
2. Benke, G.: Generalized rudin-shapiro systems. J. Fourier Anal. Appl. **1**(1), 87–101 (1994)
3. Bertalmìo, M., Sapiro, G., Caselles, V., Ballester, C.: Image inpainting. In: ACM SIGGRAPH, Association for Computing Machinery. pp. 417–424 (2000)
4. Bronstein, M., Bronstein, A., Zibulevsky, M., Zeevi, Y.: Blind deconvolution of images using optimal sparse representations. IEEE Trans. Image Process. **14**(6), 726–736 (2005)
5. Candès, E.J., Romberg, J., Tao, T.: Robust uncertainty principles: exact signal reconstruction from highly incomplete frequency information. IEEE Trans. Inform. Theor. **52**(2), 489–509 (2006)
6. Chartrand, R., Staneva, V.: Restricted isometry properties and nonconvex compressive sensing. Inverse Probl. **24**(3), 1–14 (2008)
7. Donoho, D.L.: Compressed sensing. IEEE Trans. Inform. Theor. **52**(4), 1289–1306 (2006)
8. Donoho, D.L., Elad, M.: Optimally sparse representation in general (nonorthogonal) dictionaries via ℓ_1 minimization. Proc. Nat. Acad. Sci. USA **100**(5), 2197–2202 (2003)
9. Elad, M., Aharon, M.: Image denoising via sparse and redundant representations over learned dictionaries. IEEE Trans. Image Process. **15**(12), 3736–3745 (2006)
10. Elad, M., Figueiredo, M., Ma, Y.: On the role of sparse and redundant representations in image processing. Proc. IEEE **98**(6), 972–982 (2010)
11. Elad, M., Milanfar, P., Rubinstein, R.: Analysis versus synthesis in signal priors. Inverse Probl. **3**(3), 947–968 (2007)
12. Hawe, S., Kleinsteuber, M., Diepold, K.: Cartoon-like image reconstruction via constrained ℓ_p-minimization. In: IEEE International Conference on Acoustics, Speech, and Signal Processing, Institute of Electrical and Electronics Engineers pp. 717–720 (2012)
13. Lustig, M., Donoho, D., Pauly, J.M.: Sparse MRI: The application of compressed sensing for rapid MR imaging. Mag. Reson. Med. **58**(6), 1182–1195 (2007)
14. Nam, S., Davies, M., Elad, M., Gribonval, R.: Cosparse analysis modeling – uniqueness and algorithms. In: IEEE International Conference on Acoustics, Speech and Signal Processing, Institute of Electrical and Electronics Engineers pp. 5804–5807 (2011)
15. Nocedal, J., Wright, S.J.: Numerical Optimization, 2nd edn. Springer, New York (2006)

16. Romberg, J.: Imaging via compressive sampling. IEEE Signal Process. Mag. **25**(2), 14–20 (2008)
17. Rudin, L.I., Osher, S., Fatemi, E.: Nonlinear total variation based noise removal algorithms. Physica D **60**(1–4), 259–268 (1992)
18. Selesnick, I.W., Figueiredo, M.A.T.: Signal restoration with overcomplete wavelet transforms: Comparison of analysis and synthesis priors. In: Proceedings of SPIE Wavelets XIII, The International Society for Optical Engineering. (2009)
19. Tibshirani, R., Saunders, M., Rosset, S., Zhu, J., Knight, K.: Sparsity and smoothness via the fused lasso. J. Roy. Statist. Soc. Ser. B **67**(1), 91–108 (2005)
20. Tropp, J.A., Wright, S.J.: Computational methods for sparse solution of linear inverse problems. Proc. IEEE **98**(6), 948–958 (2010)

Exploiting Structural Consistencies with Stacked Conditional Random Fields

Peter Kluegl, Martin Toepfer, Florian Lemmerich, Andreas Hotho, and Frank Puppe

Abstract Conditional Random Fields (CRF) are popular methods for labeling unstructured or textual data. Like many machine learning approaches, these undirected graphical models assume the instances to be independently distributed. However, in real-world applications data is grouped in a natural way, e.g., by its creation context. The instances in each group often share additional structural consistencies. This paper proposes a domain-independent method for exploiting these consistencies by combining two CRFs in a stacked learning framework. We apply rule learning collectively on the predictions of an initial CRF for one context to acquire descriptions of its specific properties. Then, we utilize these descriptions as dynamic and high quality features in an additional (stacked) CRF. The presented approach is evaluated with a real-world dataset for the segmentation of references and achieves a significant reduction of the labeling error.

Keywords Collective information extraction • Crf • Stacked graphical models • Structural consistencies • Rule learning

P. Kluegl (✉)
Department of Computer Science VI, University of Wuerzburg, Am Hubland, Wuerzburg, Germany

Comprehensive Heart Failure Center, University of Wuerzburg, Straubmhlweg 2a, Wuerzburg, Germany
e-mail: pkluegl@informatik.uni-wuerzburg.de

M. Toepfer • F. Lemmerich • A. Hotho • F. Puppe
Department of Computer Science VI, University of Wuerzburg, Am Hubland, Wuerzburg, Germany
e-mail: toepfer@informatik.uni-wuerzburg.de; lemmerich@informatik.uni-wuerzburg.de; hotho@informatik.uni-wuerzburg.de; puppe@informatik.uni-wuerzburg.de

P. Latorre Carmona et al. (eds.), *Mathematical Methodologies in Pattern Recognition and Machine Learning*, Springer Proceedings in Mathematics & Statistics 30, DOI 10.1007/978-1-4614-5076-4_8,

1 Introduction

The vast availability of unstructured and textual data increased the interest in automatic sequence labeling and content extraction methods over the last years. One of the most popular techniques is Conditional Random Fields (CRF) and their chain structured variant, linear chain CRFs. CRFs model conditional probabilities with undirected graphs and are trained in a supervised fashion to discriminate label sequences. Although CRFs and related methods achieve remarkable results, there remain many possibilities to increase their accuracy.

One aspect of improvement has been the relaxation of the assumption that the instances are independent and identically distributed. Relational and nonlocal dependencies of instances or interesting entities have been in the focus of collective information extraction. Due to the fact that these dependencies need to be represented in the model structure, approximate inference techniques like Gibbs Sampling or Belief Propagation are applied [7, 16]. They achieved significant improvements, but at the cost of a computationally expensive inference. It has been shown by several approaches that combined models based only on local features and exact inference can match the results of complex models while still being efficient. Kou and Cohen [10] used stacked graphical learning to aggregate the output of a base learner and to add additional features based on related instances to a stacked model. Another example is Krishnan and Manning [11] who exploit label consistencies with a two-stage CRF. However, all these approaches take only similar tokens or related instances into account while the consistencies of the structure are ignored.

Semi-structured text like any other data is always created in a certain context. This may often lead to consistencies in this creation context. While the instances are locally homogeneously distributed in one context, the dataset is globally still heterogeneous and the structure of information is possibly conflicting. The bibliographic section of a scientific publication, for example, applies a single style guide and its instances (references) share a very similar structure, while their structure might differ for different styles. Previously published approaches, cf., [8] represent structural properties directly in a higher-order model and thus suffer from a computationally expensive inference and furthermore apply a domain-dependent model.

In this paper, we propose a novel and domain-independent method for exploiting structural consistencies in textual data by combining two linear chain CRFs in a stacked learning framework with a novel intermediate step. After the instances are initially labeled, a rule learning method is applied on label transitions within one creation context in order to identify their shared properties. The stacked CRF is then supplemented with high-quality features that help to resolve possible ambiguities in the data. We evaluate our approach with a real-world dataset for the segmentation of references, a domain that is widely used to assess the performance of information extraction techniques. The results show a significant reduction of the labeling error and confirm the benefit of additional features induced online during processing the data.

The rest of the paper is structured as follows: First, Sect. 2 gives a short introduction in the background of the applied techniques. Next, Sect. 3 describes how structural consistencies can be exploited with stacked CRFs. The experimental results are presented and discussed in Sect. 4. Section 5 gives a short overview of the related work, and Sect. 6 concludes with a summary of the presented work.

2 Background

The presented method combines ideas of linear chain Conditional Random Fields (CRF), stacked graphical models and rule learning approaches. Thus, these techniques are outlined in this section.

2.1 Linear Chain Conditional Random Fields

Linear chain CRFs [12] are a chain structured variant of discriminative probabilistic graphical models. The chain structure fits well with sequence labeling tasks naturally reflecting the inherent structure of the data while providing efficient inference. By modeling conditional distributions, CRFs are capable of handling large numbers of possibly interdependent features.

Let $\mathbf{x}$ be a sequence of tokens $\mathbf{x} = (x_1, \ldots, x_T)$ referring to observations, e.g. the input text split into lexical units, and $\mathbf{y} = (y_1, \ldots, y_T)$ a sequence of labels assigned to the tokens. Taking $\mathbf{x}$ and $\mathbf{y}$ as arguments, let $f_1, \ldots, f_K$ be real-valued functions, called feature functions. To keep the model small, we restrict the linear chain CRF to be of Markov order one, i.e. the feature functions have the form $f_i(\mathbf{x}, \mathbf{y}) = \sum_t f_i(\mathbf{x}, y_{t-1}, y_t, t)$. A linear chain CRF of Markov order one has K model parameters $\lambda_1, \ldots, \lambda_K \in \mathbb{R}$, one for each feature function, by which it assigns the conditional probability

$$P_\lambda(\mathbf{y}|\mathbf{x}) = \frac{1}{Z_\mathbf{x}} \exp\left(\sum_{t=1}^{T}\sum_{i=1}^{K} \lambda_i \cdot f_i(y_{t-1}, y_t, \mathbf{x}, t)\right),$$

to $\mathbf{y}$ with a given observation $\mathbf{x}$.

The feature functions typically indicate certain properties of the input, e.g. capitalization or the presence of numbers, while the model parameters weight the impact of the feature functions on the inference. The partition function $Z_\mathbf{x}$ normalizes $P_\lambda(\mathbf{y}|\mathbf{x})$ by summing over all possible label sequences for $\mathbf{x}$. The properties of a token x_t indicated by feature functions usually consider a small fixed sized window around x_t for a given state transition. In the following, we will use the terms feature and feature function interchangeably.

2.2 Stacked Graphical Learning

Stacked Graphical Learning is a general meta learning algorithm, cf., [10]. First, the data is processed by a base learner based on conventional features representing local characteristics. Subsequently, every single data instance is expanded by information about the inferred labels of related instances. In a second stage, a stacked learner exploits this extra information. The process of aggregating and projecting the predicted information on instance features to support another stacked learner can be repeated several times. Stacking graphical models has two central advantages: The approach enables to model long-range dependencies among related instances and the inference for each learner remains effective.

Similar to Wolpert's Stacked Generalization [17], Stacked Graphical Models use a cross-fold technique in order to avoid overfitting. As a result, the stacked learners get to know realistic errors of their input components that would also occur during runtime. However, Stacked Generalization and Stacked Graphical Models are essentially different approaches. In short, Stacked Generalization learns a stacked learner to combine the output of several different base learners on a per instance basis. In contrast, Stacked Graphical Learning utilize a stacked learner to aggregate and combine the output of one base learner on several instances, thus supporting collective inference.

2.3 Rule Learning

In this paper, we propose to utilize rule-based descriptions as an intermediate step of our general approach, cf. Sect. 3. For this task we will transfer the data into a tabular form of attribute value pairs and learn rules on this data representation. While over the last decades a large amount of rule learning approaches have been proposed, we will concentrate on two main approaches in this paper:

Ripper [4] is probably the most popular learning algorithm for learning a *set* of rules. Ripper learns rules one at a time by growing and pruning each rule and then adds them to a result set until a stop criterion is met. After adding a rule to the result, examples covered by this rule are then removed from the training data. Ripper is known to be on par regarding classification performance with other learning algorithms for rule sets, e.g., C 4.5, but is computationally more efficient.

As an alternative, we utilize Subgroup Discovery [9] (also called Supervised Descriptive Rule Discovery or Pattern Mining) to describe structural consistencies. In this approach, an exhaustive search for the best k conjunctive patterns in the dataset with respect to a pre-specified target concept and a quality function, e.g., the F_1 measure, is performed. Additionally different constraints on the resulting patterns can be applied, e.g., on the maximum number of describing attribute value pairs or the minimum support for a rule. While the resulting rules are not intended to be used directly as a classifier, a related approach using patterns based on improvement of the target share and additional constraints has recently been successfully applied as an intermediate feature construction step for classification tasks [2].

3 Method

For introducing the proposed method, we first motivate the problem. Then, the stacked inference, the induction of the structural properties, and the parameter estimation are presented.

3.1 Problem Description

Recap the inference formula of CRFs (cf. Sect. 2.1). From the model designers' perspective, the classification process is mainly influenced by the choice of the feature functions f_i. The feature functions need to provide valuable information to discriminate labels for all possible kinds of instances. This works well when the feature functions encode properties that have the same meaning for inference across arbitrary instances. For example in the domain of reference segmentation, some special words have a strong indicative meaning for a certain task: The word identity feature "WORD=proceedings" always suggests labeling the token as "Booktitle." Thus, the learning algorithm will fix the corresponding weights to high values, leading the inference procedure into the right direction. Some features, however, violate the assumption of a consistent meaning. Their validity depends on a special context or is restricted through long-range dependencies. In our example of reference extraction, the feature that indicates colons might suggest an author label if the document finishes author fields with colons. However, other style guides define a different structure of the author labels. Consequently, the learning algorithm assigns the weights to average the overall meaning. On the one hand, this yields good generalization given enough training data. On the other hand, averaging the weights of such features restricts them to stay behind their discriminative potentials. If we knew that a certain feature has a special meaning inside the given context, we could do better by increasing (or decreasing) the weights, dynamically adapting to the given context. This procedure cannot be performed independently of the remaining weights. Hence, we apply a different approach in this paper. Instead of changing the model parameters, we learn the weights of additional feature functions describing the structural and context-dependent consistencies.

3.2 Stacked Inference

Sequence labeling methods like CRFs assign a sequence of labels $\mathbf{y} = (y_1, \ldots, y_T)$ to a given sequence of observed tokens $\mathbf{x} = (x_1, \ldots, x_T)$. Let *crf* $(\mathbf{x}, \Lambda, F) = \mathbf{y}$ be the function that applies the CRF model with the weights $\Lambda = \{\lambda_1, \ldots, \lambda_K\}$ and the set of feature functions $F = \{f_1, \ldots, f_K\}$ on the input sequence $\mathbf{x}$ and returns the labeling result $\mathbf{y}$. The set of model weights must of course correspond to the set of feature functions. Since the CRF processes this sequence of tokens in one labeling task, we call $\mathbf{x}$ an instance. All instances together form the dataset D which is split in a disjoint training and testing subset. An information or entity often consists of several tokens and is encoded by a sequence of equal labels. We assume here that the given labels already specify an unambiguous encoding. An instance itself may contain multiple entities specified by an arbitrary amount of labels, one label for each token of the input sequences. Furthermore, we assume that the dataset $D = \{C_1, \ldots, C_n\}$ can be completely and disjointly partitioned into subsets of instances $\mathbf{x}$ that originate from the same creation context C_i. Similar to the relational template in [10], we imply that a trivial context template exists for the assignment of the context set. Staying with the previous examples, the reference section of this paper defines a context C with 18 instances.

In stacked graphical learning, several models can be stacked in a sequence. Experimental results, e.g., of Kou [10], have shown that this approach already converges with a depth of two learners and no significant improvements are achieved with more iterations of stacking. Therefore, we only apply stacked graphical learning with CRFs in a two-stage approach like Krishnan and Manning [11]. In order to extract entities collectively, we define the stacked inference task on the complete set of instances $\mathbf{x}$ in one context C. The two CRFs, however, label the single instances within that context separately as usual. The following algorithm summarizes the stacked inference combined with online rule learning. Section 3.3 describes the rule learning techniques for the identification of structural consistencies and how the "meta-features" f^m are induced. Details about the estimation of the weights (e.g., $\hat{\Lambda}$ or Λ^m) are discussed in Sect. 3.4.

1. *Apply base CRF*
   Apply $crf\ (\mathbf{x}, \hat{\Lambda}, F) = \hat{\mathbf{y}}$ on all instances $\mathbf{x} \in C$ in order to create the initial label sequences $\hat{\mathbf{y}}$.
2. *Learn structural consistencies*
   Learn classification rules for certain label transitions of all instances $\mathbf{x} \in C$ and construct a feature function $f^m \in F^m$ for each discovered rule.
3. *Apply stacked CRF*
   Apply $crf\ (\mathbf{x}, \Lambda \cup \Lambda^m, F \cup F^m) = \mathbf{y}$ again on all instances $\mathbf{x} \in C$ in order to create the final label sequences $\mathbf{y}$.

3.3 Learning Structural Consistencies During Inference

First, the overall idea how structural consistencies are learned during the inference is addressed. The technical details are then described after a short example.

We apply rule learning techniques on all (probably erroneous) label assignments $\hat{\mathbf{y}} \in C$ of the base CRF. The rules are learned in order to classify certain label transitions and, thus, describe the shared properties of the transition within the context C. The labeling error in the input data is usually eliminated by the

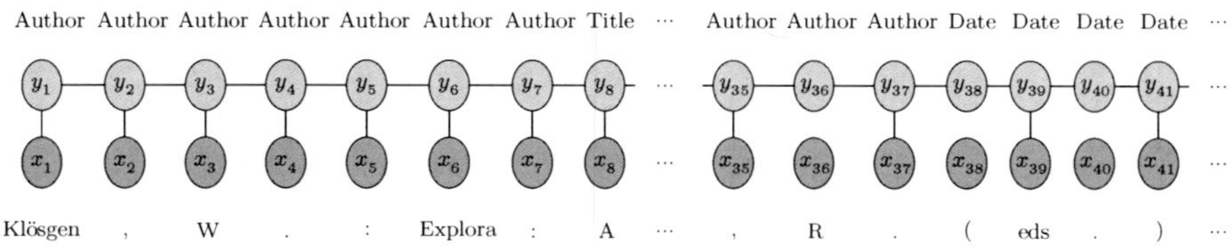

Fig. 1 Two excerpts of the ninth reference with erroneous labeling: The begin of the title (token x_6 and x_7) was falsely labeled as author. The editor was additionally labeled as an author (up to token x_{37}) and date (token x_{38} to x_{41})

generalization of the rule learning algorithm. The label transition is optimally described by a single pattern that covers the majority of transitions despite of erroneous outliers. The learned rules are then used as binary feature functions in the same context C: They return 1 if the rule applies on the observed token x_t, and 0 otherwise. We gain additional features that indicate label transitions if the instances are consistently structured. Even if the learned rules are misleading due to erroneous input data or missing consistency of the instances, their discriminative impact on the inference is yet to be weighted by the learning algorithm of the stacked CRF.

This process is illustrated by a simple example concerning the author label, but can also be applied to any other label. Let the reference section of this paper be processed by the base CRF that classified all instances but one correctly. For some reasons the base CRF misclassified tokens x_6 and x_7 and tokens x_{15} to x_{41} in the ninth reference (cf. Fig. 1). The input of the rule learning now consists of 18 transitions from author to title whereas one transition is incorrect. In this case, a reasonable result of the rule learning is the rule "if the token x_t is a colon after a period, then there is a transition from author to title at t." Converted to a feature function, this rule returns 1 for token x_5 and 0 for all other tokens of the reference in Fig. 1. Therefore, the stacked CRFs' likelihood of a transition from author to title is increased at token x_5 due to high weights of the meta-features. Furthermore, no meta-features for the transition from author to date could be learned resulting in a decreased evidence for the label sequence at token x_{37}.

In general, any classification method can be applied to learn indicators for the structural consistencies. In this work, we restrict ourselves to techniques for supervised descriptive rule discovery because their learning and inference algorithm are efficient and the resulting rules can be interpreted. This allows studies about the properties of good descriptions of structural consistencies. We disregarded the usage of the Support Vector Machines [5] because several models need to be trained and executed during the stacked inference.

For inducing the meta-features, a tabular database $T = (I, A)$ is created for each context C as the input of the rule learning techniques described in Sect. 2.3. The database is constructed using all instances $\mathbf{x} \in C$, their corresponding initially labeled sequences $\hat{\mathbf{y}}$ and a feature set $F' \subseteq F$. Each individual of I corresponds to a single token of the instances in C. The set of attributes A consists of the possible labels and a superset of F': When classification methods are applied on sequential data, the attributes are also added for a fixed window, e.g., the attribute "WORD@-1=proceedings" indicates that the token before the current individual equals the string "proceedings." Hence, this superset contains the feature functions F and additionally their manifestation in a window defined by the window size w. The cells in the tabular database T are filled with binary values. They are set to true if the feature or label occurs at this token and to false otherwise.

In a next step, the target attributes for the rule learning are specified. In this work, we apply the transition of two different labels. Here, the target attribute is set on all transitions of two dedicated labels in the initially labeled result $\hat{\mathbf{y}}$ of the context C. Finally, the set of learned rules is transformed to the set of binary feature functions F^m that return true, if the condition of the respective rule applies.

3.4 Parameter Estimation

The weights of two models need to be estimated for the presented approach: the parameters of the base model ($\hat{\Lambda}$) and of the stacked model ($\Lambda \cup \Lambda^m$). The base model needs to be applied on the training instances for the estimation of the weights of the stacked model, i.e., step 1 and step 2 of the stacked inference in Sect. 3.2 need to be performed on the training set. If the weights of the base model are estimated as usual using the labeled training instances, then it produces unrealistic predictions on these instances and the meta-features of the stacked model are over-fitted resulting in a decrease of accuracy. Since in this case the base model is optimized on the training instances, it labels these instances perfectly. The learned rules create optimal descriptions of the structural consistencies and the stacked model assigns biased weights to the meta-features. This is of course not reproducible when processing unseen data.

The simple solution to this problem is a cross-fold training of the base model for the training of the stacked CRF as described in Sect. 2.2 and successfully applied by several approaches [10, 11]. Training of the base model in a cross-fold fashion is also a very good solution for the presented approach. However, we simply decrease the accuracy of the model by reducing the training iterations for estimating Λ'. Thus, only one model needs to be trained for the learning phase of the stacked model. For the testing phase or common application on the other hand, a single base model learned with the default settings is applied.

The model of the stacked CRF is trained dependent on the base model and the creation context C. Both are applied to induce the new features online during the stacked inference. These meta-features possess the same meaning in the complete dataset, but change their interpretation or manifestation depending on the currently processed creation context. The weights $\Lambda = \{\lambda_1, \ldots, \lambda_K\}$ and $\Lambda^m = \{\lambda_1^m, \ldots, \lambda_M^m\}$ of the stacked CRF are estimated to maximize the conditional probability on the instances of the training dataset:

$$P_\lambda\left(\mathbf{y}|\mathbf{x},C,crf\left(\mathbf{x},\Lambda',F\right)\right) = \frac{1}{Z_\mathbf{x}}\exp\left(\sum_{t=1}^{T}\sum_{i=1}^{K}\lambda_i \cdot f_i(y_{t-1},y_t,\mathbf{x},t)\right.$$
$$\left. + \sum_{t=1}^{T}\sum_{j=1}^{M}\lambda_j^m \cdot f_j^m\left(y_{t-1},y_t,\mathbf{x},t,C,crf\left(\mathbf{x},\Lambda',F\right)\right)\right)$$

The resulting model still relies on the normal features functions but is extended with dynamic and high-quality features that help to resolve ambiguities and substitute for other missing features.

A short example: The induced feature function for the transition of the author to the title is set to very high weights for the corresponding state transition of the learned model. As illustrated in the example of Sect. 3.3, this feature function returns 1 in the reference section of this paper for a token which is a colon after a period. In other reference sections with a different style guide applied, the feature function for this state transition returns 1, if the token is a period and is followed by a parenthesis. However, both examples refer only to exactly one feature function that dynamically adapts to the currently processed context.

4 Experimental Results

The presented approach is evaluated in the domain of reference segmentation. The common approach is to separately process the instances, namely the references. Within these references, the interesting entities need to be identified. Since all tokens of a reference are part of exactly one entity, one speaks of a segmentation task. In this section, we introduce the overall settings and present the experimental results.

4.1 Datasets

All available and commonly used datasets for the segmentation of references are a listing of references without their creation context and are thus not applicable for the evaluation of the presented approach. Therefore, a new dataset was manually annotated with the label set of Peng and McCallum [14] concerning the fields *Author*, *Booktitle*, *Date*, *Editor*, *Institution*, *Journal*, *Location*, *Note*, *Pages*, *Publisher*, *Tech*, *Title* and *Volume*. The resulting dataset contains 566 references in 23 documents extracted only of complete reference sections of real publications. The amount of instances is comparable to previously published evaluations in this domain, cf. [6, 14].

Two different sets of features are used in the experimental study: The basic features are applied for all evaluated models and correspond to the features of well-known evaluations in this domain, cf. [6,14]. For an extensive definition of the set of basic features, we refer to the dataset that contains all applied basic features. Only a part of the basic features is used for the induction of the meta-features, omitting ngram and token window features. This restriction is justified with their minimal expressiveness for the identification of the structure in relation to the increase of the search space. The dataset with all applied basic features can be freely downloaded.[1]

4.2 Implementation Details

The machine learning toolkit Mallet[2] is used for an implementation of the CRF in the presented approach. For rule learning, we integrated two different methods. We chose a subgroup discovery implementation[3] because of the multifaceted configuration options that allow a deep study of the approach's limits. Additionally, we applied an established association rule learner Ripper [4] for a comparable implemention.[4]

We used only the default parameters for the CRF and all evaluated models were trained until convergence. Only for the training of the stacked model, the iterations of the base model were reduced to 50 iterations. For the default configuration of both rule learning tasks, we set the window size $w = 1$. Additionally for the default setting of the subgroup discovery, we used a quality function based on the F_1 measure, selected only one rule for each description of a label, restricted the length of the rules to maximal three selectors, and set an overall minimum threshold of the quality of a rule equal to 0.5.

4.3 Performance Measure

The performance is measured with commonly used methods of the domain. Let tp be the number of true positive classified tokens and define fn and fp, respectively, for false negatives and false positives. Since punctuations contain no target information in this domain, only alpha-numeric tokens are considered. *Precision*, *recall* and F_1 are computed by:

$$\text{precision} = \frac{tp}{tp+fp}, \quad \text{recall} = \frac{tp}{tp+fn}, \quad F_1 = \frac{2 \cdot \text{precision} \cdot \text{recall}}{\text{precision} + \text{recall}}.$$

[1] http://www.is.informatik.uni-wuerzburg.de/staff/kluegl_peter/research/

[2] http://mallet.cs.umass.edu

[3] http://sourceforge.net/projects/vikamine/

[4] http://sourceforge.net/projects/weka/

4.4 Results

The presented approach is compared to two baseline models in a five-fold cross evaluation. Four different settings of stacked CRFs combined with a rule learning technique are investigated. A detailed description of all evaluated models is given in Table 1. The documents of the dataset are randomly distributed over the fivefolds.

The results of the experimental study are depicted in Table 2. Only marginal differences can be observed between the two baseline models CRF and STACKED CRF. This indicates that the normal stacking approach cannot exploit the structural consistencies or gain much advantage of the predicted labels.

All of our stacked models combined with rule learning techniques significantly outperform the baseline models using one-sided, paired t-tests on the F_1 scores of the single references ($p \ll 0.01$). Comparing the results of STACKED+DESCRIPTIVE that only considers the consistencies of four labels to the baseline CRF, our approach achieves an average error reduction of over 30% on the real-world dataset.

The lower F_1 scores of STACKED+RIPPER can be explained by its learning algorithm. The Ripper implementation applies a coverage-based learning in order to create a set of rules, which together classify the target attribute. This can lead to a reproduction of errors of the predicted labels in the description of the structure. In the domain of reference segmentation a single description of the structure is preferable. However, in other domains where disjoint consistencies of one transition can be found, a covering algorithm for inducing the rules performs probably better.

The second configuration STACKED+MORE considers the transition between seven labels and is able to slightly increase the measured F_1 score compared to our default model STACKED+DESCRIPTIVE. STACKED+MAX that induces rules for all labels achieves only an average error reduction of 26% compared to a single CRF. This is mainly caused by misleading meta-features for rare labels. The task of learning consistencies from a minimal amount of examples is error-prone and can decrease the accuracy, especially if the examples are labeled incorrectly.

Table 1 Overview of the evaluated models

CRF	A single CRF trained on the same data and features.
STACKED CRF	A two-stage CRF approach. The predictions of the base CRF are added as features to the stacked CRF.
STACKED+DESCRIPTIVE	The default approach of stacked CRF combined with subgroup discovery for rule learning. Only transitions between the labels *Author*, *Title*, *Date*, and *Pages* that commonly occur in most references are considered.
STACKED+RIPPER	A stacked CRF combined with the association rule learner Ripper. Only the four most common labels are addressed.
STACKED+MORE	A stacked approach using subgroup discovery that additionally learns the transitions of the labels *Booktitle*, *Journal*, and *Volume*.
STACKED+MAX	A stacked approach using subgroup discovery that considers the transitions of all labels for the rule learning task.

Table 2 F_1 scores averaged over the fivefolds

	Average F_1
Base line	
CRF	91.3
STACKED CRF	91.8
Our approach	
STACKED+DESCRIPTIVE	94.0
STACKED+RIPPER	93.6
Variants of DESCRIPTIVE	
STACKED+MORE	94.1
STACKED+MAX	93.6

Table 3 F_1 scores of the author label

	CRF	STACKED+ DESCRIPTIVE	Error reduction
Fold 1	97.7	99.6	82.6%
Fold 2	97.0	99.2	73.3%
Fold 3	96.4	96.5	2.8%
Fold 4	97.1	98.8	58.6%
Fold 5	89.5	95.1	53.3%
Average	95.5	97.8	51.6%

Table 3 provides closer insights into the benefit of the presented approach using the author label as an example. STACKED+DESCRIPTIVE is able is significantly improve the labeling accuracy for all folds but one. The third fold contains an unfavorable distribution of style guides between the training and testing set for the author. If the initial base CRF labels a label systematically incorrectly, then the rule learning cannot induce any valuable and correct descriptions of the structure. Nevertheless, an average error reduction of over 50% is achieved for identifying the author of the reference.

To our knowledge, no domain-independent approach was published that can be utilized for a comparable model. As comparison, we applied the skip-chain approach of [16] with factors for capitalized words and additionally for identical punctuation marks, but no improvement over the baseline models could be measured. Furthermore, the feature induction for CRFs [13] was integrated, but resulted counter-intuitively in a decrease of the accuracy.

The performance time of the presented approach for onefold averaged over the fivefolds is several times faster than a higher-order model with skip edges, about nine times faster using the subgroup discovery and about fourteen times faster using Ripper. The difference in speed is less compared to previously published evaluations [10]. This is mainly caused by the fact that the rule learning is optimized neither for this task nor for the domain, e.g., by pruning the attributes.

5 Related Work

In the following, we give a brief overview on related work coming from different domains with context consistencies, attempts utilizing complex graphical models and stacked graphical models for collective information extraction, and approaches on feature induction.

Especially for Named Entity Recognition (NER) modeling long-distance dependencies is crucial. The labeling of an entity is quite consistent within a given document, however, conclusive discriminating features are sparsely spread across the document. As a consequence, leveraging predictions of one instance to disambiguate others is essential. Bunescu et al. [3], Sutton et al. [16], and Finkel et al. [7] extended the commonly applied linear chain CRF to higher order structures. The exponential increase in model complexity enforces to switch from exact to approximate inference techniques. Stacked graphical models [10, 11] retain exact inference as well as efficiency by using linear chain CRF.

In Kou and Cohen's Stacked Graphical Learning framework [10], information is propagated by Relational Templates C. Although each C may find related instances and aggregate their labels in a possibly complex manner, they utilize rather simple aggregators, e.g. COUNT and EXISTS. Likewise, the approach of Krishnan and Manning [11] uses straightforward but for NER efficient aggregate features, concentrating on the label predictions of the same entity in other instances. In contrast to Stacked Graphical Learning, they also include corpus-level features, aggregating predictions across documents. In this paper, we use data mining techniques to determine rich context sensitively applied features. Rather than simply transferring labels of related instances, e.g., by majority vote aggregation, we exploit structural properties of a given context. We represent the gathered context knowledge by several meta-features which are conceptually independent of the label types.

A semi-supervised approach on exploiting structural consistencies of documents has been taken by Arnold and Cohen [1] who improve domain adaption by conditional frequency information of the unlabeled data. They show that differences in the frequency distribution of tokens across different sections in biological research papers can provide useful information to extract protein names. Counting frequencies can be done efficiently and the experimental results suggest that these features are robust across documents. However, in general unlabeled data is not enough to model the context structure, e.g., frequency information can be noisy or differences in the frequency distribution may be caused non-structural. We propose to mine the distributions of predicted labels and their combinations with observed features to capture context structure.

Yang et al. use structural consistencies for information extraction from web forums [18]. They employ Markov Logic Networks [15] with formulas to encode the assumed structural properties of a typical forum page, e.g., characteristic link structures or tag and attribute similarities among different posts and sites. Since context structure is represented inside the graphical model, inference and learning have to fight model complexity. Another example for content extraction from web sites that exploits related instances is Gulhane et al. [8]. They assume two properties of web information: The values of an attribute distributed over different pages are similar for equal entities and the pages of one web site share a similar structure due to the creation template. In contrast to those two approaches, the work presented in this paper relies on no structural knowledge previously known about the domain.

McCallum contributed an improvement for CRF applications through feature induction [13]. Based on a given training set useful combinations of features are computed, reducing the number of model parameters. The feature induction of our approach is performed online during processing the document and applies flexible data mining techniques to specify the properties of consistent label transitions.

6 Conclusions

We have presented a novel approach for collective information extraction using a combination of two CRFs together with rule learning techniques to induce new features during inference. The initial results of the first CRF are exploited to gain information about the structural consistencies. Then, the second CRF is automatically adapted to the previously unknown composition of the entities. This is achieved by changing the manifestation of its features dependent on the currently processed set of instances. To our best knowledge, no similar and domain-independent approach was published that is able to exploit the structural consistencies in textual data. The results on a real-world dataset for the segmentation of references indicate a significant improvement towards the commonly applied models. This is achieved without any additional domain knowledge, integrated matching methods with a bibliographic database, or other jointly performed tasks like entity resolution.

References

1. Arnold, A., Cohen, W.W.: Intra-document structural frequency features for semi-supervised domain adaptation. In: Proceeding of the 17th ACM Conference on Information and Knowledge Management, pp. 1291–1300. ACM, New York (2008)
2. Batal, I., Hauskrecht, M.: Constructing classification features using minimal predictive patterns. In: Proceedings of the 19th ACM International Conference on Information and Knowledge Management, pp. 869–878. CIKM '10, ACM, New York (2010)
3. Bunescu, R., Mooney, R.J.: Collective information extraction with relational markov networks. In: Proceedings of the 42nd Annual Meeting on Association for Computational Linguistics. ACL '04, Association for Computational Linguistics, Stroudsburg, PA (2004)
4. Cohen, W.W.: Fast effective rule induction. In: Proceedings of the Twelfth International Conference on Machine Learning, pp. 115–123. Morgan Kaufmann, Los Altos (1995)
5. Cortes, C., Vapnik, V.: Support-vector networks. Mach. Learn. **20**(3), 273–297 (1995)
6. Councill, I., Giles, C.L., Kan, M.Y.: ParsCit: an open-source CRF reference string parsing package. In: Proceedings of the Sixth International Language Resources and Evaluation (LREC'08). ELRA, Marrakech, Morocco (2008)

7. Finkel, J.R., Grenager, T., Manning, C.: Incorporating non-local information into information extraction systems by gibbs sampling. In: Proceedings of the 43rd Annual Meeting on Association for Computational Linguistics, pp. 363–370. ACL '05, Association for Computational Linguistics, Stroudsburg, PA (2005)
8. Gulhane, P., Rastogi, R., Sengamedu, S.H., Tengli, A.: Exploiting content redundancy for web information extraction. Proc. VLDB Endow. **3**, 578–587 (2010)
9. Klösgen, W.: Explora: a multipattern and multistrategy discovery assistant. In: Fayyad, U., Piatetsky-Shapiro, G., Smyth, P., Uthurusamy, R. (eds.) Advances in Knowledge Discovery and Data Mining. Menlo Park, CA, USA, pp. 249–271. (1996)
10. Kou, Z., Cohen, W.W.: Stacked graphical models for efficient inference in markov random fields. In: Proceedings of the 2007 SIAM International Conference on Data Mining, SIAM, address of publisher is not known. probably. Minneapolis, Minnesota, USA (2007)
11. Krishnan, V., Manning, C.D.: An effective two-stage model for exploiting non-local dependencies in named entity recognition. In: Proceedings of the 21st International Conference on Computational Linguistics and the 44th Annual Meeting of the ACL, pp. 1121–1128. ACL-44, ACL, Stroudsburg, PA (2006)
12. Lafferty, J., McCallum, A., Pereira, F.: Conditional random fields: probabilistic models for segmenting and labeling sequence data. In: Proceedings of the 18th International Conference on Machine Learning, Morgan Kaufmann Publishers Inc., San Francisco, CA, USA, pp. 282–289 (2001)
13. McCallum, A.: Efficiently inducing features of conditional random fields. In: Nineteenth Conference on Uncertainty in Artificial Intelligence (UAI03), Morgan Kaufmann Publishers Inc., San Francisco, CA, USA (2003)
14. Peng, F., McCallum, A.: Accurate information extraction from research papers using conditional random fields. In: HLT-NAACL, Association for Computational Linguistics, Boston, Massachusetts, USA, pp. 329–336 (2004)
15. Richardson, M., Domingos, P.: Markov logic networks. Mach. Learn. **62**(1–2), 107–136 (2006)
16. Sutton, C., McCallum, A.: Collective segmentation and labeling of distant entities in information extraction. In: ICML Workshop on Statistical Relational Learning and Its Connections to Other Fields (2004)
17. Wolpert, D.H.: Stacked generalization. Neural Networks **5**, 241–259 (1992)
18. Yang, J.M., Cai, R., Wang, Y., Zhu, J., Zhang, L., Ma, W.Y.: Incorporating site-level knowledge to extract structured data from web forums. In: Proceedings of the 18th International Conference on World Wide Web, pp. 181–190. ACM, New York (2009)

Detecting Mean-Reverted Patterns in Algorithmic Pairs Trading

K. Triantafyllopoulos and S. Han

Abstract This paper proposes a methodology for detecting mean-reverted segments of data streams in algorithmic pairs trading. Considering a state-space model that describes the spread (data stream) as the difference of the prices of two assets, we propose two new recursive least squares (RLS) algorithms for predicting mean-reversion of the spread in real time. The first is a combination of steepest descent RLS and Gauss–Newton RLS, for which we extend previous work by providing exact recursive equations to update the variable forgetting factor (VFF). We propose a new RLS algorithm for variable forgetting, by transforming the prediction errors into a binary process and adopting Bayesian methods for inference. The new approach is versatile as compared to more traditional RLS schemes, having the advantage of uncertainty analysis around the VFF. The methods are illustrated with real data, consisting of daily prices of Target Corporation and Walmart Stores Inc shares, over a period of 6 years. Alongside the detection of mean-reversion of the spread, we implement a simple trading strategy. The empirical results suggest that the new Bayesian approach returns are in excess of 130 % cumulative profit over a period of 2 years.

Keywords Pairs trading • Statistical arbitrage • Mean-reversion • Market-neutral trading • Recursive least squares • Variable forgetting factor • Adaptive filtering

1 Introduction

This paper is concerned with the detection of mean-reversion in algorithmic pairs trading. Pairs trading is a market-neutral trading philosophy, which exploits a very basic trading rule in the stock market: buy low and short-sell high. This and similar

---

K. Triantafyllopoulos (✉) • S. Han
School of Mathematics and Statistics, University of Sheffield, Sheffield S3 7RH, UK
e-mail: k.triantafyllopoulos@sheffield.ac.uk; s.han@sheffield.ac.uk

P. Latorre Carmona et al. (eds.), *Mathematical Methodologies in Pattern Recognition and Machine Learning*, Springer Proceedings in Mathematics & Statistics 30,
DOI 10.1007/978-1-4614-5076-4_9,

ideas can be traced back in the 1930s [1], but they appeared within the pairs trading framework in the 1980s with the work of Nunzio Tartaglia and his quantitative group at Morgan Stanley. Algorithmic pairs trading deploys trading strategies and related financial decisions that can be implemented in the computer without human intervention. Recently, there has been a growing interest in pairs trading and in related market-neutral trading approaches, see, e.g., [2, 3, 7, 8, 14]; Gatev et al. [3] provide a nice historical introduction to the subject. For book-length discussion of pairs trading, the reader is referred to [13] and [9].

Considering the spread of two assets A and B, defined as the difference of the prices of A and B, pairs trading assumes that the spread attains an equilibrium or that the spread in the long run reverts to its historical mean. The main idea behind pairs trading is to propose trades based upon the relative temporary mispricings of the two assets. For example, suppose that the equilibrium of the spread is \$10 (US dollars) and today the two assets trade at \$40 and \$10, respectively, or with spread \$40 − \$10 = \$30. Then, pairs trading suggests to go short (or short-sell) asset A (as this is likely to be overpriced at \$40) and to go long (or buy) asset B (as this is likely to be underpriced at \$10). If the spread reverts to its historical mean, the price of asset A will decrease and/or the price of asset B will increase, either of which can return a healthy profit.

This approach is heavily dependent on the assumption of mean-reversion for the spread. Should this assumption be violated, the trader may buy an overpriced asset, which is losing its value, or may short-sell an undervalued asset, which commit the trader to high buying costs in the future; both of these actions result in significant loss. Mean-reversion implies that the spread fluctuates around the equilibrium level and thus if today the price of an asset goes up, it will go down in the near future and vice versa. Conversely, a breakdown of mean-reversion implies that any shock in the spread may be permanent and hence there is no guarantee that if today the price of an asset goes up, it will go down in the near future. This is what happened at the Wall-Street operating Long-Term Capital Management hedge fund, which had to be bailed out in 1998 by the Federal Reserve Bank of New York over a \$3.625 billion loss, of which \$286 million was in equity pairs [5]. This story reveals that spread speculation, in particular regarding to short-selling assets, may lead to significant loss, if mean-reversion is not monitored systematically and if the uncertainty of spread prediction is not studied carefully. In practice, assets may exhibit local mean-reversion, i.e. there may be periods of mean reversion followed by periods of a breakdown of mean-reversion, see, e.g., [9, 12]. As a result, it is proposed that by detecting periods of mean reversion, the trader can find opportunities for trading.

In this paper, considering a state-space model for the spread, we propose that mean-reverted patterns can be detected in real time by predicting the dynamic states of this model. Section 2 motivates and describes the above model. In Sect. 3 we propose a new Gauss–Newton recursive least squares (RLS) method, which improves on the Gauss–Newton RLS algorithm of [11]. The proposed variable forgetting updating is exact, avoiding the approximation of the second derivative of the mean square error [11]. By combining in a single algorithm, steepest descent RLS [4, 6] and Gauss–Newton RLS, we propose that for periods when the data stream is smooth or stable, the former algorithm is used, while for higher adaptivity tracking at periods when the data stream is not stable, the algorithm switches to the Gauss–Newton RLS. Furthermore, Sect. 3 introduces a new variable forgetting factor RLS algorithm, based on Bayesian conjugate methods. According to this the forgetting factor is stochastic and, under some assumptions, we derive its predictive distribution. This approach of variable forgetting is versatile, as it enables the modeller to analyze uncertainty associated with the forgetting factor at each point of time. In Sect. 4 a simple trading strategy is proposed, which is put into practice in the empirical section that follows. Section 5 illustrates the proposed methodology considering a data stream consisting of share prices of Target Corporation and Walmart Stores Inc, over a period of 6 years. The empirical findings suggest that the proposed Bayesian methods outperform both the RLS and the new combined RLS algorithm. Section 6 gives concluding remarks and the appendix details proofs of arguments of Sect. 3.

2 Model Set-Up

Suppose that, at time t, $p_{\mathrm{A},t}$ denotes the price of asset A and $p_{\mathrm{B},t}$ denotes the price of asset B, and that we form the data stream $\{y_t\}$, where $y_t = p_{\mathrm{A},t} - p_{\mathrm{B},t}$. Assuming that in the long run, $\{y_t\}$ is mean-reverted, we can take advantage of temporary mispricings of A and B, in order to realize profits. For example, if $\{y_t\}$ reverts to a mean 10, and at the current time point t, $y_t = 18$, this would mean that asset A is overpriced and/or asset B underpriced; hence, a simple trading strategy is to go short (short-sell) asset A and go long (buy) asset B, with the view to realize a profit when y_{t+1} reverts to its mean 10. On the other hand, if the spread $\{y_t\}$ is not mean-reverted, it is quite dangerous to deploy the above trading procedure as we can short-sell an undervalued asset or buy an overvalued asset, both cases of which leading to a loss.

A model that can detect mean-reverted patterns is important as it will enable decision makers to quantify risks and construct optimal trading strategies. A first such model is to consider that y_t is a noisy version of a mean-reverted process x_t and that the associated noise can make or break the mean-reversion. This model is technically defined by $y_t = x_t + \xi_t$ and $x_t = \alpha + \beta x_{t-1} + \zeta_t$, where α and β are parameters and ξ_t, ζ_t are uncorrelated white noise processes [2]. With this model in place it is easy to observe that the mean of y_t converges to $\alpha/(1-\beta)$, if and only if $|\beta| < 1$, where $|\beta|$ is the absolute value of β. A disadvantage of the above model is that the parameters α and β are time-invariant, which may not be a realistic assumption as the mean-reverting behaviour of $\{y_t\}$ may well change over time. An improvement of this model is achieved by considering the more general state-space model

$$y_t = \alpha_t + \beta_t y_{t-1} + \epsilon_t = F_t^{\mathrm{T}} \theta_t + \epsilon_t, \tag{1}$$

$$\theta_t = \Phi \theta_{t-1} + \omega_t, \tag{2}$$

where now the parameters α_t and β_t are time-varying [12]. This is conveyed via the Markovian bivariate column vector $\theta_t = (\alpha_t, \beta_t)^{\mathrm{T}}$ and $F_t = (1, y_{t-1})^{\mathrm{T}}$, where T denotes transposition. The matrix Φ is assumed to be diagonal, i.e. $\Phi = \mathrm{diag}(\phi_1, \phi_2)$, for some known ϕ_1 and ϕ_2 (in the applications of this paper we consider $\phi_1 = \phi_2 \approx 1$; the above authors describe maximum likelihood estimation of these hyperparameters). Furthermore, the white noise ω_t is assumed to be uncorrelated of ϵ_t, for all t. The system is initiated at $t = 1$ with θ_1. The distributions of ϵ_t, ω_t and θ_1 may be assumed as Gaussian, and this is assumed for the rest of the paper.

With model (1)–(2) in place, the condition of mean reversion is $|\beta_t| < 1$, for all t; for a proof of this result the reader is referred to [12]. Furthermore, we suggest that one can extract a mean-reverted segment $I_{t_1,t_2} = [t_1, t_2] \subsetneq [1, N]$, where $t_1 < t_2$ and N is the total length of $\{y_t\}$. In other words, we propose that using data up to t, we obtain a working prediction $\hat{\beta}_{t+1}$ of β_{t+1}, and we use this to test whether $t+1$ belongs to a mean-reverted segment $t+1 \in [t_1, t_2]$ (with $|\hat{\beta}_{t_1}| < 1, \ldots, |\hat{\beta}_{t_2}| < 1$) or not; usually we operate in real-time one-step ahead, i.e. $t_1 = t_2 = t+1$. For this process to end and in order to declare at each time point whether the process is likely to be mean-reverted, the $\hat{\beta}_t$ values are needed. In a similar fashion we require to obtain predictions of y_{t+1}, which are denoted by $\hat{y}_{t+1}$. Adaptive schemes of both of these predictions are described next.

3 Adaptive Learning

Given data $D_t = (y_1, \ldots, y_t)$, we wish to predict β_{t+1} and y_{t+1}. In the sequel we present three prediction procedures with adaptive forgetting: (a) a standard steepest descent variable forgetting factor RLS (SDvFF) algorithm, (b) a new switching variable forgetting factor RLS (SvFF), which switches from SDvFF to a modified Gauss–Newton method and (c) a new adaptive forgetting method using a binary model (BBvFF).

3.1 Steepest Descent Variable Forgetting Factor RLS

RLS algorithms with adaptive memory have been extensively developed and used [4], while SDvFF with adaptive memory is recently introduced in [6]. Defining λ to be a forgetting factor ($0 < \lambda \leq 1$) (so that the system forgets observations in a rate $\sum_{j=0}^{\infty} \lambda^j y_{t-j} = (1-\lambda)^{-1}$), then locally λ is chosen so that to minimize the cost function $J(t) = 2^{-1} E(e_t^2)$, where $E(\cdot)$ denotes expectation and $e_t = y_t - \hat{y}_t$ is the residual or prediction error at time t. Using the celebrated Kalman filter for the recursions of $m_t = E(\theta_t | D_t)$, we update λ_t via

$$\lambda_t = [\lambda_{t-1} - a\nabla_\lambda(t)]_{\lambda_-}^{\lambda_+}, \tag{3}$$

where $\nabla_{\lambda}(t) = \partial J(t)/\partial\lambda$ is the first derivative of $J(t)$ with respect to λ, a is the rate of change from $t-1$ to t and the notation $y = [x]_a^b$ implies that $y = x$, if $a < x < b$, $y = a$, if $x \leq a$ and $y = b$, if $x \geq b$. This consideration is important, as it makes sure that $\lambda_1, \dots, \lambda_t, \dots$ do not fall below λ_- or above λ_+, both of which could result in poor or unacceptable values for the forgetting factor. The values of a, λ_- and λ_+ are pre-specified and usually they are chosen by considering historical data [6].

The algorithm starts by first applying the Kalman filter recursions (all dependent on λ_{t-1}), e.g. the prediction of θ_{t+1} is $\hat{\theta}_{t+1} = \Phi m_t$ and $m_t = \Phi m_{t-1} + K_t e_t$, where $K_t = (\lambda_{t-1} + F_t^{\mathrm{T}} \Phi P_{t-1} \Phi^{\mathrm{T}} F_t)^{-1} \Phi P_{t-1} \Phi^{\mathrm{T}} F_t$ is the Kalman gain, with P_{t-1} being the covariance matrix of θ_{t-1}, which is updated with a similar formula.

Then the algorithm commences by deriving a formula for the evaluation of $\nabla_{\lambda}(t)$, which is used in (3). We note that $\nabla_{\lambda}(t) \approx -e_t F_t^{\mathrm{T}} \Phi \psi_{t-1}$, where $\psi_t = \partial m_t / \partial \lambda$ and then the recursion of ψ_t is given by $\psi_t = (I - K_t F_t^{\mathrm{T}}) \Phi \psi_{t-1} + S_t F_t e_t$, where I denotes the identity matrix and $S_t = \partial P_t / \partial \lambda$. After some algebra, we can verify that S_t can be recursively computed by

$$S_t = -\lambda_t^{-1} P_t + \lambda_t^{-1} K_t K_t^{\mathrm{T}} + \lambda_t^{-1} (I - K_t F_t^{\mathrm{T}}) \Phi S_{t-1} \Phi^{\mathrm{T}} (I - F_t K_t^{\mathrm{T}}).$$

With this algorithm in place we obtain a sequence of estimated values $\{m_t, P_t, \psi_t, S_t, \lambda_t\}$, given that some initial values $m_1, P_1, \psi_1, S_1, \lambda_1$ are specified. The prediction $\hat{\beta}_{t+1}$, which is needed to detect stationary patterns in the spread, is then obtained by noting $\theta_t = (\alpha_t, \beta_t)^{\mathrm{T}}$, i.e. $\hat{\beta}_{t+1}$ is the lower element of the bivariate vector Φm_t. Furthermore, one can obtain the prediction $\hat{y}_{t+1}$ of the spread y_{t+1}, as $\hat{y}_{t+1} = F_t^{\mathrm{T}} \Phi m_t$.

3.2 *Switch Variable Forgetting Factor RLS*

The SDvFF algorithm described above works well when there are small shifts in the data stream; some authors assume that $\{y_t\}$ is a stable or stationary data stream, see, e.g. [11]. However, as this assumption may well not be valid—and indeed in the context of this paper we are interested in detecting patterns of stable or mean-reverted regions—we need to have a way that the forgetting factor will adapt much quicker to changes than the steepest descent method of the previous section.

Motivated by this observation, we deploy a Gauss–Newton method, for the recurrence updating of the adaptive forgetting factor λ_t. Thus, we replace (3), by

$$\lambda_t = \left[\lambda_{t-1} - a \frac{\nabla_{\lambda}(t)}{\nabla_{\lambda}^2(t)} \right]_{\lambda_-}^{\lambda_+}, \tag{4}$$

where $\nabla_{\lambda}^2(t)$ denotes the second partial derivative of $J(t)$ with respect to λ.

The Gauss–Newton algorithm, abbreviated as GN, follows similar lines as the SDvFF algorithm of the previous section. The recursions of m_t, P_t, ψ_t, S_t are as before, but now we need to add the evaluation of $\nabla_{\lambda}^2(t)$. By direct differentiation, we get $\nabla_\lambda^2(t) \approx (F_t^{\mathrm{T}}\Phi\psi_{t-1})^2 - e_t F_t^{\mathrm{T}}\Phi\eta_t$, where $\eta_t = \partial\psi_{t-1}/\partial\lambda$. In the appendix it is shown that by applying differentiation, the recursions of η_t and $L_t = \partial S_t/\partial\lambda$ are

$$\eta_t = (I - K_t F_t^{\mathrm{T}})\Phi\eta_{t-1} + L_t F_t e_t - 2S_t F_t F_t^{\mathrm{T}}\Phi\psi_{t-1}, \tag{5}$$

where

$$\begin{aligned} L_t = {} & \lambda_t^{-1}(I - K_t F_t^{\mathrm{T}})\Phi L_{t-1}\Phi^{\mathrm{T}}(I - F_t K_t^{\mathrm{T}}) + \lambda_t^{-2}P_t(I - F_t K_t^{\mathrm{T}}) - \lambda_t^{-1}S_t + M_t + M_t^{\mathrm{T}} \\ & -\lambda_t^{-2}(I - K_t F_t^{\mathrm{T}})\Phi S_{t-1}\Phi^{\mathrm{T}}(I - F_t K_t^{\mathrm{T}}) \end{aligned} \tag{6}$$

and

$$M_t = \lambda_t^{-1}S_t F_t F_t^{\mathrm{T}}\{P_t - \Phi S_{t-1}\Phi^{\mathrm{T}}(I - F_t K_t^{\mathrm{T}})\}. \tag{7}$$

In this algorithm, the estimated vectors $\{\hat{\theta}_t\}$ are provided by the sequences of $\{m_t\}$ and $\{P_t\}$, but now these depend on the forgetting factor λ_t, which is updated by the Gauss–Newton method (4), for the application of which the sequences $\{\psi_t, S_t, L_t, M_t\}$ need to be computed.

However, experience with data indicates that this algorithm is too sensitive to abrupt changes, due to the introduction of the second derivative in (4). This results in poor predictive performance, in particular in the case of non-stationary abrupt points of the data stream being followed by smooth less noisy data points. In theory, the SDvFF algorithm of the previous section is likely to work better when there are not many abrupt changes (providing more smooth predictions of β_t), while the GN algorithm is likely to work better when there are abrupt changes that require the forgetting factor to adapt quickly. Motivated from this observation, we propose that the forgetting factor is updated as follows

$$\lambda_t = \begin{cases} [\lambda_{t-1} - a\nabla_\lambda(t)]_{\lambda-}^{\lambda+}, & \text{if} \quad |e_t| \le k_t \\ \left[\lambda_{t-1} - a\frac{\nabla_\lambda(t)}{\nabla_\lambda^2(t)}\right]_{\lambda-}^{\lambda+}, & \text{if} \quad |e_t| > k_t \end{cases}$$

where $|\cdot|$ indicates modulus and k_t is a pre-specified threshold value.

Basically this scheme suggests a combination of SDvFF and GN methods, which switches from SDvFF to GN when the prediction errors are high (SvFF). According to this, at each time t, if the prediction error e_t is small, then we operate with SDvFF, because we consider that the system performs smoothly. If, however, there is an abrupt change—e.g. as evidence of an outlier present or evidence of non-stationarity of the stream—then we operate with the Gauss–Newton updating of the forgetting factor. This allows us to experience smooth or low variance predictions when the data stream is not noisy and adaptive predictions when the stream is noisy.

3.3 Beta-Bernoulli Variable Forgetting Factor RLS

Motivated by the comments in the last paragraph, it is desirable to obtain adaptive forgetting to respond quickly to changes of the data stream, but yet retaining smooth performance when the data is not as noisy. A natural way to achieve this is to set the forgetting factor λ as

$$\lambda = \pi\lambda_+ + (1-\pi)\lambda_-, \tag{8}$$

where λ_-, λ_+ are defined as before and π is the probability that the prediction error e_t is small (to be defined mathematically later). This setting guarantees that $\lambda_- \leq \lambda \leq \lambda_+$ and so there is no need to force λ to be in this range (as we had to do in (3) and (4)). The motivation of formula (8) is that when e_t^2 is small, then π should be close to one and so λ is much closer to λ_+, while when e_t^2 is large (e.g. in the presence of an outlier), then π should be close to zero and λ is closer to λ_-. In the sequel, we propose a flexible mechanism for estimating π from the data and thus learning about λ from the data.

In order to estimate π we deploy conjugate Bayesian methods. We define x_t to be a binary random variable (taking values 0 and 1), according to the law

$$x_t = \begin{cases} 1, \text{ if } & |e_t| \leq k_t, \quad \text{with probability } \pi \\ 0, \text{ if } & |e_t| > k_t, \quad \text{with probability } 1-\pi \end{cases}$$

where the probability π is unknown and k_t is a threshold determined at time t. Given π, the likelihood function from the observed data x_t is a bernoulli distribution, written as $p(x_t \mid \pi) = \pi^{x_t}(1-\pi)^{1-x_t}$.

The next step is to specify a prior distribution for π. Since π is a probability, a natural choice is a beta distribution, i.e. $\pi \sim \text{Be}(c_1, c_2)$, with density

$$p(\pi) = \frac{\Gamma(c_1+c_2)}{\Gamma(c_1)\Gamma(c_2)}\pi^{c_1-1}(1-\pi)^{c_2-1},$$

where $c_1, c_2 > 0$ are the parameters of the beta distribution and $\Gamma(\cdot)$ denotes the gamma function. The reader should note that the distribution of π is implicitly conditional on data up to time $t-1$ (i.e. c_1, c_2 will depend on $t-1$ as shown below). Then by applying Bayes theorem, we have that the posterior distribution of π, given x_t is

$$p(\pi \mid x_t) \propto p(x_t \mid \pi)p(\pi) \propto \pi^{c_1+x_t-1}(1-\pi)^{c_2+1-x_t-1},$$

which is proportional to the posterior beta distribution, $\pi \mid x_t \sim \text{Be}(c_1+x_t, c_2-x_t+1)$. Applying this formula sequentially we obtain $\pi \mid x_1, \ldots, x_t \equiv \pi \mid D_t \sim \text{Be}(c_{1t}, c_{2t})$, where $c_{1t} = c_{1,t-1} + x_t$ and $c_{2t} = c_{2,t-1} - x_t + 1$. With these results in place, we propose that after observing y_t, the mean of the distribution $\pi \mid D_t$ as a prediction of π and so we write $\hat{\pi}_t = E(\pi \mid D_t) = c_{1t}(c_{1t}+c_{2t})^{-1}$. We can then propose the variable forgetting factor $\hat{\lambda}_t$ as the mean of $\lambda \mid D_t$, i.e. $\hat{\lambda}_t = E(\lambda_t \mid D_t) = \hat{\pi}_t\lambda_+ + (1-\hat{\pi}_t)\lambda_-$.

The above procedure is delivered conditional on the specification of the threshold k_t. Since the innovations $\{\epsilon_t\}$ and $\{\omega_t\}$ are assumed to be Gaussian, we have that $q_t^{-1/2}e_t$ follows the standard Gaussian distribution and so $P(q_t^{-1/2}|e_t| \leq 1.96) = 0.95$, where $q_t = \text{Var}(e_t \mid D_{t-1}) = F_t^T P_{t-1} F_t / \hat{\lambda}_{t-1} + 1$ is the prediction error variance, and so we can choose $k_t = 1.96 q_t^{1/2}$.

In this framework, the forgetting factor is a random variable (as it is a function of π) and thus, by rewriting (8) as $\lambda = (\lambda_+ - \lambda_-)\pi + \lambda_-$ and noting the posterior beta distribution of $\pi \mid D_t$, one can derive the distribution of $\lambda \mid D_t$, i.e.

$$p(\lambda \mid D_t) = c(\lambda_t - \lambda_-)^{c_{1t}-1}(\lambda_+ - \lambda_t)^{c_{2t}-1}, \tag{9}$$

where the proportionality constant c is given by

$$c = \frac{\Gamma(c_{1t} + c_{2t})}{(\lambda_+ - \lambda_-)^{c_{1t}+c_{2t}-1}\Gamma(c_{1t})\Gamma(c_{2t})}.$$

This distribution is consistent with the above evaluation of $\hat{\lambda}_t = E(\lambda \mid Dt)$, but $p(\lambda \mid D_t)$ offers versatility, as one can analyze the uncertainty associated with $\hat{\lambda}_t$. In particular, the variance of λ is

$$\text{Var}(\lambda \mid D_t) = \frac{c_{1t}c_{2t}(\lambda_+ - \lambda_-)}{(c_{1t} + c_{2t})^2(c_{1t} + c_{2t} + 1)}.$$

This can be easily verified by first writing $\lambda_t = (\lambda_+ - \lambda_-)\pi + \lambda_-$, so that $\text{Var}(\lambda \mid D_t) = (\lambda_+ - \lambda_-)^2 \text{Var}(\pi \mid D_t)$ and then noting the variance of the beta distribution $\pi \mid D_t \sim \text{Be}(c_{1t}, c_{2t})$. Furthermore, for $c_{1t} > 1$ and $c_{2t} > 1$, the mode of λ_t is

$$\text{mode}(\lambda \mid D_t) = \frac{\lambda_+(c_{1t} - 1) + \lambda_-(c_{2t} - 1)}{c_{1t} + c_{2t} - 2}, \tag{10}$$

the proof of which is detailed in the appendix.

Given initial values $c_{1,1}$ and $c_{2,1}$, the above development suggests a sequential algorithm, which basically runs the Kalman filter conditional on the forgetting factor λ_{t-1} and then updates the forgetting factor according to the above beta-bernoulli procedure. From the prior distribution $\pi \sim \text{Be}(c_{1,1}, c_{2,1})$, by noting that $\hat{\pi}_1 = c_{1,1}(c_{1,1} + c_{2,1})^{-1}$, we propose to set the initial values as $c_{1,1} = c_{2,1} = 0.5$. This is motivated by the reasoning that since we have no data observed, we predict the probability π as 0.5.

Although (9) provides the distribution of λ, in the empirical section below, $\hat{\lambda}_t$ is used as the working estimate of λ. It is then of interest to derive a recursive formula of $\hat{\lambda}_t$ (as a function of $\hat{\lambda}_{t-1}$), e.g. to enable comparison with the forgetting factors of SS-RLS and GN-RLS algorithms. To this end, we note that $\hat{\pi}_t = c_{1t}(c_{1t} + c_{2t})^{-1}$ and so

$$\hat{\lambda}_t = \hat{\pi}_t\lambda_+ + (1-\hat{\pi}_t)\lambda_- = \frac{c_{1t}\lambda_+ + c_{2t}\lambda_-}{c_{1t}+c_{2t}}$$
$$= \frac{c_{1,t-1}+c_{2,t-1}}{c_{1t}+c_{2t}}\hat{\lambda}_{t-1} + \frac{(\lambda_+ - \lambda_-)x_t + \lambda_-}{c_{1t}+c_{2t}}$$
$$= \frac{t-1}{t}\hat{\lambda}_{t-1} + \frac{(\lambda_+ - \lambda_-)x_t + \lambda_-}{t}, \tag{11}$$

upon observing that $c_{1t}+c_{2t} = c_{1,t-1}+c_{2,t-1}+1 = \cdots = c_{1,1}+c_{2,1} = t$, if we use the initial values $c_{1,1} = c_{2,1} = 0.5$, suggested above.

An interesting implication of equation (11) is as follows. Rewrite (11) as $t\hat{\lambda}_t = (t-1)\hat{\lambda}_{t-1} + (\lambda_+ - \lambda_-)x_t + \lambda_-$ and apply this formula sequentially backwards in time to obtain $\hat{\lambda}_t = t^{-1}(0.5\lambda_+ + (t-0.5)\lambda_-) + t^{-1}(\lambda_+ - \lambda_-)\sum_{j=1}^{t} x_j$. Then for large t we have that the first term of the above equation is close to λ_- and so $\hat{\lambda}_t \approx \lambda_- + (\lambda_+ - \lambda_-)\bar{x}$, which can be rearranged to

$$\hat{\lambda}_t \approx \bar{x}\lambda_+ + (1-\bar{x})\lambda_-, \tag{12}$$

where $\bar{x} = t^{-1}\sum_{j=1}^{t} x_i$ is the mean of the x_t values. Equation (12) has a similar form as that of λ, where now $\bar{x}$ replaces π. The former is the proportion of points $(1,\ldots,t)$ for which the model has a good forecast ability (judged via $\Pr(|e_t| \le k_t)$).

In the application of the above procedure, a difficulty is associated with the phenomenon of having many units in the x_t sequence (corresponding to small errors $|e_t|$, followed by a small number of outlying observations or zeros (corresponding to large value of $|e_t|$). Since, in the evaluation of $\hat{\lambda}_t$, we estimate π by $\bar{x}$, it is possible to have one or more sequential outliers, but the probability $\hat{\pi}$ can be large, as it is estimated by $\bar{x}$ (which includes many units from past observations). This may be fine in the long run, as it can be argued that units are much more dominant than zeros, in this particular case. But in the short run it can cause the forgetting factor to fail to adapt as new information comes in. To illustrate this point, suppose that $\lambda_- = 0.8$, $\lambda_+ = 0.99$ and say that we have got the first 95 time points corresponding to $|e_t| \le k_t$ $(t = 1,\ldots,95)$ and for the last 5 time points $(t = 96,\ldots,100)$, it is $|e_t| > k_t$. This implies that $x_t = 1$, for $t = 1,\ldots,95$ and $x_t = 0$, for $t = 96,\ldots,100$, so that $\bar{x} = 0.95$. Therefore, at time $t = 100$, the forgetting factor is $\hat{\lambda}_t = 0.9805$, which is much closer to $\lambda_+ = 0.99$ than to $\lambda_- = 0.8$ and thus it is failing to be adaptive after 5 consecutive outliers.

To alleviate this problem, we propose a simple intervention step. Basically, we suggest that once a change in the sequence x_t occurs we reset the prior values of $c_{1,t-1}$ and $c_{2,t-1}$ to the initial values $(c_{1,1} = c_{2,1} = 0.5)$ and so we reset the prior $\hat{\pi}_{t-1}$ to 0.5. This simple intervention has the effect of not using $\bar{x}$ in (12), but only the current value, or more mathematically

$$\hat{\pi}_t = \frac{c_{1t}}{c_{1t}+c_{2t}} = \frac{c_{1,1}+x_t}{c_{1,1}+c_{2,1}+1} = \frac{1+2x_t}{4},$$

where $x_t = 1$ (if $x_{t-1} = 0$) and $x_t = 0$ (if $x_{t-1} = 1$). As a result the forgetting factor at time t is

$$\hat{\lambda}_t = \begin{cases} 0.75\lambda_+ + 0.25\lambda_-, & \text{if } x_t = 1 \\ 0.25\lambda_+ + 0.75\lambda-, & \text{if } x_t = 0 \end{cases}$$

In the numerical example above, we would intervene after the first time we have observed an outlier, that is at $t = 96$, with $\hat{\lambda}_{96} = 0.25 \times 0.99 + 0.75 \times 0.8 = 0.847$. This forgetting factor is much closer to 0.8 than to 0.9805 (which was the value obtained above using $\bar{x}$) and it depicts the influence of the outlier (and the proposed intervention) at this point of time.

4 Trading Strategy

In this section we describe a simple trading strategy, which is used in the empirical section below. As mentioned previously, we have two assets A and B and at each time t, we observe their prices, denoted as $p_{A,t}$ and $p_{B,t}$. Consequently, their spread is formed as $y_t = p_{A,t} - p_{B,t}$. If we knew y_{t+1} at time t, we would know which asset to buy and which to sell. If say, we knew that $y_t < y_{t+1}$, we would know that the price of A was likely to increase at $t+1$ and / or the price of B was likely to decrease at $t+1$ (relative to their prices at t). As a result at time t, we should buy asset A and short-sell B. At $t+1$, with $y_{t+1} > y_t$, we would realize a profit, if we sold asset A and bought B (minus transaction costs). Of course at time t we do not know y_{t+1}, but we can forecast it by $\hat{y}_{t+1}$ (produced by each of the three algorithms).

Suppose that we wish to open a trading position at time t. We first check whether the spread is expected to be mean-reverted at $t+1$, i.e. we see whether $|\hat{\beta}_{t+1}| < 1$. If $|\hat{\beta}_{t+1}| \geq 1$, we decide not to trade and so we do not open a position at t. If $|\hat{\beta}_{t+1}| < 1$, we open a trading position according to the rule: buy a unit of A and short-sell 3 units of B, if $\hat{y}_{t+1} - h > y_t$ and short-sell 3 units of A and buy a unit of B, if $\hat{y}_{t+1} + h < y_t$. Here $h > 0$ is a prediction margin that allows some uncertainty to guarantee that the unknown y_{t+1} at time t falls in the range $[\hat{y}_{t+1} - h, \hat{y}_{t+1} + h]$. To explain further this idea, imagine that at time t, the spread is equal to $y_t = 10$ and that we project that at time $t+1$ the spread prediction goes up to $\hat{y}_{t+1} = 11$. As there is uncertainty around this prediction, it is equally likely that the true value of y_{t+1} be 12 (higher than y_t) or 9 (lower than y_t), each of which returns a different trading rule (buy/sell or sell/buy); in particular the latter ($y_{t+1} = 9 < y_t$) can result in a loss, if we implement the rule $y_t > \hat{y}_{t+1}$. For this reason, introducing h prevents this happening. In this simple example, if we operate with h as 10% of $\hat{y}_{t+1} = 1.1$, then $\hat{y}_{t+1} - k = 9.9 < 10 = y_t$ and so we will not open the position: buy A and short-sell B. Likewise $\hat{y}_{t+1} + h = 11.1 > 10 = y_t$ and so we do not open the position: short-sell A and buy B. In such a case, we make the decision not to open a trading position at t, because the predicted $\hat{y}_{t+1}$ does not create a safe margin in the spread to allow for a probable profit. Figure 1 illustrates the trading strategy we propose.

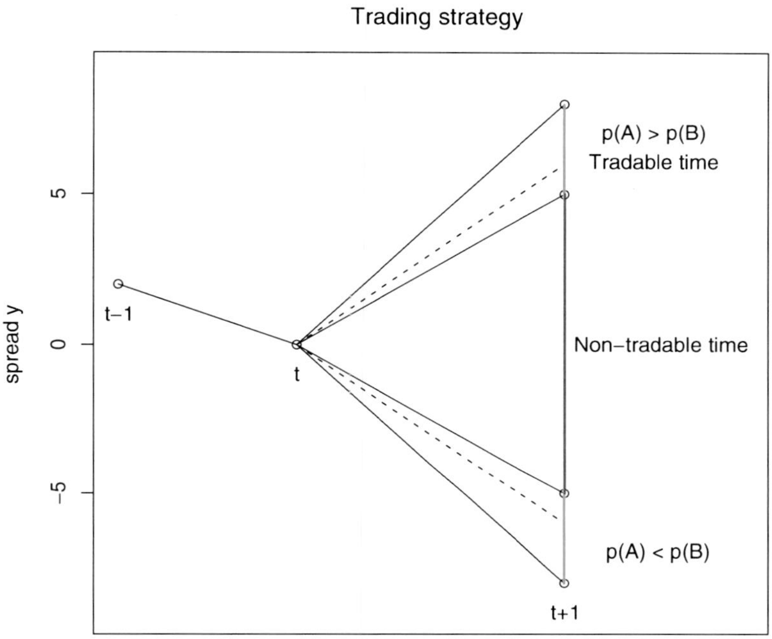

Fig. 1 Proposed trading strategy. The *dashed lines* indicate the unknown value of y_{t+1} prior to time $t+1$ and the related *solid lines* show the interval $[\hat{y}_{t+1} - h, \hat{y}_{t+h} + h]$. The plot indicates tradable and non-tradable situations according to the values of the prediction mean $\hat{y}_{t+1}$ and the prediction margin h

We can see that the lower the value of h the more transactions we operate (we are more exposed to risk) and the higher the value of h the less transactions we operate (we are more conservative). As a result, one has to evaluate h and in this paper we propose to look at the mean cumulative profit considering h as 1 %, 3 % and 5 % of $\hat{y}_{t+1}$; this is implemented in the empirical section below.

Coming back to our trading strategy, with N the length of the spread, at each time $t = 3, \ldots, N-1$, we close the trading position from time $t-1$ (if at that point of time we had opened a trade) by reverting the buy and sell actions of time $t-1$ (i.e. if we bought A and sold B at $t-1$, we sell A and we buy B at t). At t, we project whether at $t+1$ will be mean-reverted at $t+1$ and if it is, we open a trading position according to the rules above. Thus, at each time t we do one of the following: (a) close a position of $t-1$ (if we opened a position at $t-1$) and open a position at t, (b) close a position of $t-1$ (if we opened a position at $t-1$) and we decide not to open a new trade, (c) open a new trade at t, or (d) no action is committed. Initially, at $t=2$, we may open the first trading position and at the last point $t=N$, we may close the trading position of time $N-1$.

5 Example: Target and Walmart Shares

In this section we consider data, consisting of daily prices (in US$) of Target Corporation and Walmart Stores Inc share prices, over a period of nearly 6 years (3 January 2006 to 14 October 2011). The data has been provided by Yahoo! finance (http://finance.yahoo.com/). Figure 2 shows the data and its spread $y_t = p_{At} - p_{Bt}$ (shown in the inset), where p_{At} denotes the daily price of Target (for short) and p_{Bt} denotes the daily price of Walmart (for short). The figure indicates that the relative prices of these two assets seem to co-evolve at some periods of time and that their spread appears to exhibit mean-reversion, fluctuating around a historical mean of 0.7713. However, it is clear that the spread is more mean-reverting at about 2008 and after 2010, while the fluctuations of 2006 until 2008 may give some indication of lack of mean-reversion. Figure 2 also shows that Walmart is much more volatile and uncertain than Target, which seems to have relatively moderate fluctuations around its historical mean.

We have applied the three algorithms (SDvFF, SvFF and BBvFF) of the previous section, in order to obtain predictions $\hat{y}_t$ of the data stream y_t and $\hat{\beta}_t$ of the β_t coefficient, which controls the detection of mean-reversion. For the application of SDvFF and SvFF we have used a learning rate $a = 0.5$; this choice is supported by our own simulation experiments (not reported here) as well as by suggestions of [6]. For the specification of the threshold k_t in the updating of λ_t in SvFF, we apply the same approach as in BBvFF, i.e. we switch from SDvFF to Gauss–Newton when $q_t^{-1/2}|e_t| > 1.96$ or when there is an outlying observation causing the standardized one-step error $q_t^{-1/2}e_t$ to be large enough in modulus. For SDvFF and SvFF we have used $\lambda_1 = 0.95$ and for *BBvFF* we have used $\lambda_1 = 0.8$; we have found that the algorithms are not sensitive to this choice, but a lower value in BBvFF reflects a prior belief of a noisy signal. For all algorithms we have used $\lambda_+ = 0.99$. We wish to allow a low forgetting factor λ_- to be small, in order to facilitate quick adaptive performance when needed. However, initial experiments show that both SDvFF and SvFF failed to allow this and in fact for λ_- lower values than 0.9 the algorithms crashed; this observation is in agreement with similar studies in the literature, see, e.g., [4] or [6]. Thus for these two algorithms we chose $\lambda_- = 0.9$. BBvFF did not experience such a problem and thus we were able to set up a low value for λ_-; we observed that for $\lambda_- = 0.9$ its performance was in par with SvFF, but the real benefits adopting this approach were obtained when considering lower values of λ_- and thus making BBvFF more adaptive than the other two algorithms. We thus chose $\lambda_- = 0.01$ for BBvFF; for a smooth signal we operate with a relatively large forgetting factor (close to λ_+), for a noisy signal the intervention step allows the VFF to be much lower (close to $\lambda_- = 0.01$). Figure 3 shows the dynamics of the three forgetting factors. Figures 4–6 show the predictions $\hat{\beta}_t$ for each of the three algorithms together with the predictions of the spread. In each algorithm we detect mean-reversion locally if the modulus of the estimated $\hat{\beta}_t$ is less than 1.

In Fig. 4 the SDvFF algorithm has detected about a year (mid-2008–mid-2009) of breakdown in mean reversion and this is reflected in the very poor prediction of

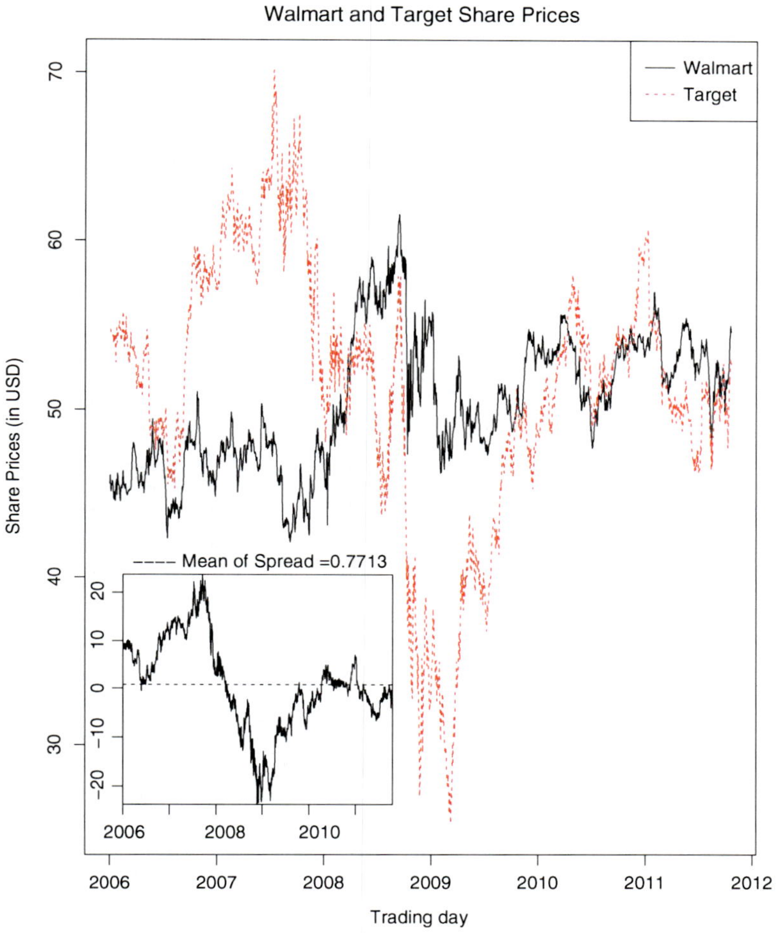

Fig. 2 Share prices of Walmart and Target together with their spread data stream (inset)

the spread in that time. Other than this, post to 2009 there is a long period of mean-reversion, while in the early/mid 2007 there is a suggestion of another breakdown of mean reversion. In comparison with Fig. 4, Fig. 5 shows an improved spread fit (the dashed line with the predictions is much closer to the solid line of the actual spread). The breakdown of mean reversion is of shorter length, which indicates a better performance of the algorithm, e.g. in 2009 there is still a breakdown of mean reversion, but now the prediction system recovers much quicker and thus it allows for some trades even at this period of high uncertainty. We observe a degree of similarity between SvFF and BBvFF, evidenced by similar periods of breakdown of mean reversions in Figs. 5 and 6, especially in the period of end 2006 till the end

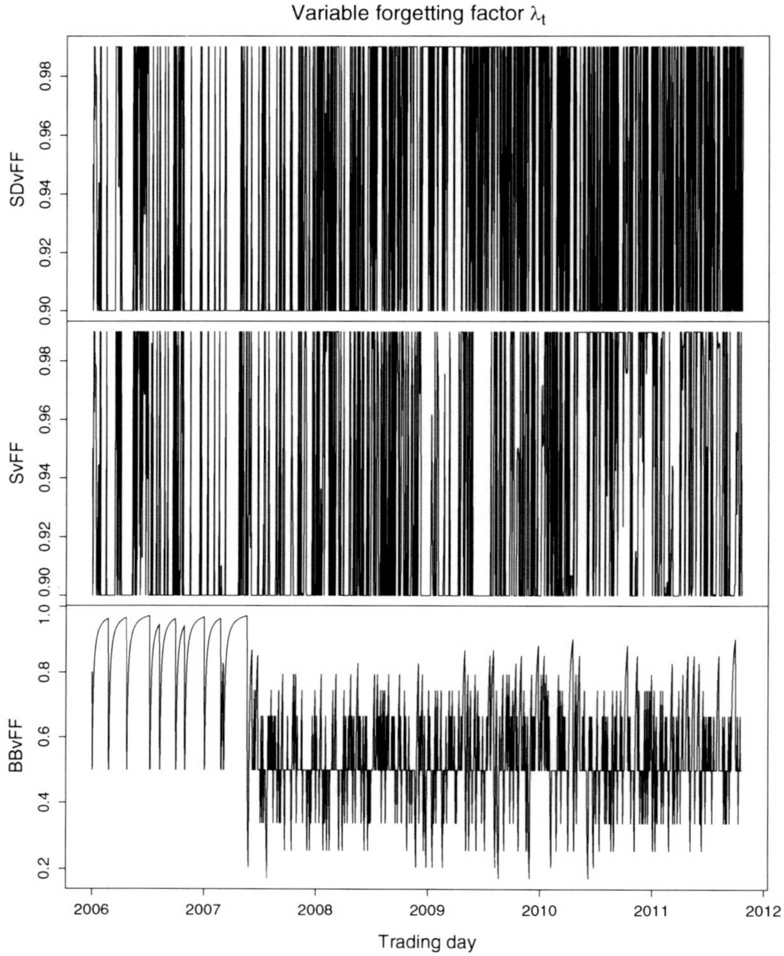

Fig. 3 Dynamics of the variable forgetting factor of each of the three algorithms (SDvFF, SvFF and BBvFF). We observe that the range of the VFF in SDvFF and SvFF is restricted in a relatively small window (0.95,0.99), while that of BBvFF has a much wider range (0.01,0.99)

of 2007. BBvFF seems to have more clear signals on breakdown of mean reversion, while SvFF seems to have more fluctuations around it; both seem to outperform SDvFF.

Figure 7 shows the mean square prediction error (MSE) computed over time, by each of the algorithms. We observe that up to the mid-2008 the three algorithms have similar MSE values, but after this time the MSE of SDvFF explodes (this clearly is due to the very poor spread predictions in that period, see Fig. 4). Up to 2010, the best performer is the SvFF algorithm, but considering the time interval 2010–2011,

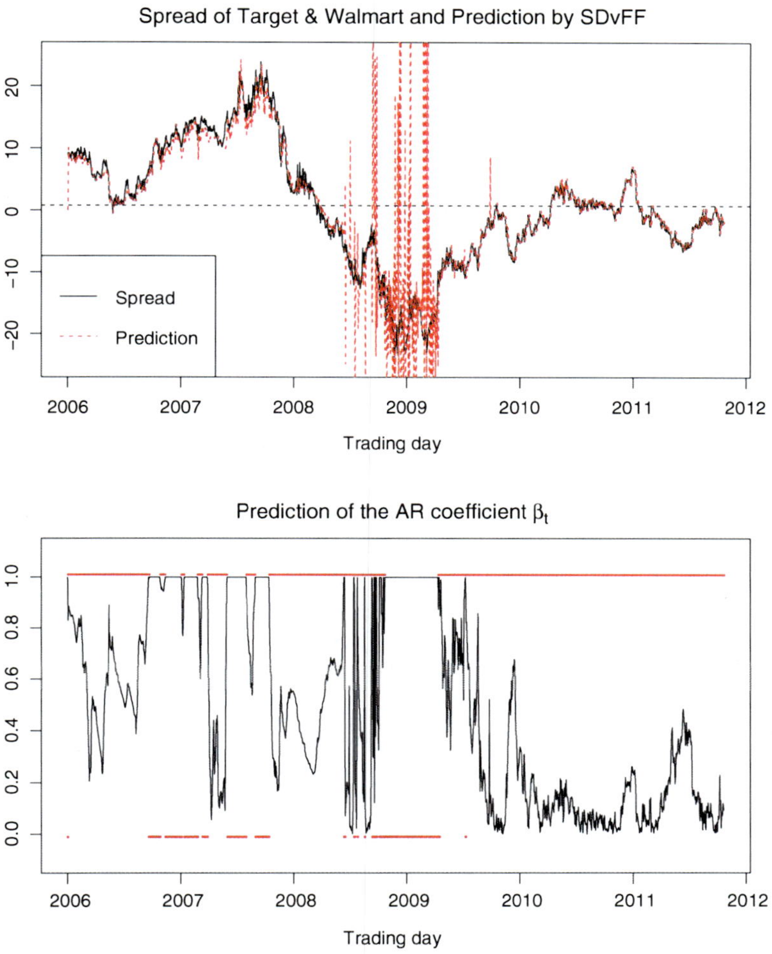

Fig. 4 Spread prediction using the SDvFF algorithm (*upper panel*) and modulus of the predicted coefficients $\{\hat{\beta}_t\}$ using the SDvFF algorithm (*lower panel*). Mean-reversion segments are detected for t satisfying $|\hat{\beta}_t| < 1$ and these regimes are highlighted with the *upper straight line* (the *lower straight line* indicates periods of breakdown of mean reversion)

BBvFF has the best performance. It also appears that when BBvFF has a poor prediction influencing the MSE, this is not as bad as in the other two algorithms (their respective MSE after a poor prediction are higher with SDvFF providing clearly the most poor performance). If one looks at the overall period or at the period 2010–2011, then BBvFF is the best performer here.

We have implemented the trading strategy, described in Sect. 4, from 4 January 2010 to 14 October 2011. Table 1 shows the mean, the standard deviation and the final closing balance of the cumulative profits for $h = 1\%, 3\%$ and 5% of $\hat{y}_{t+1}$,

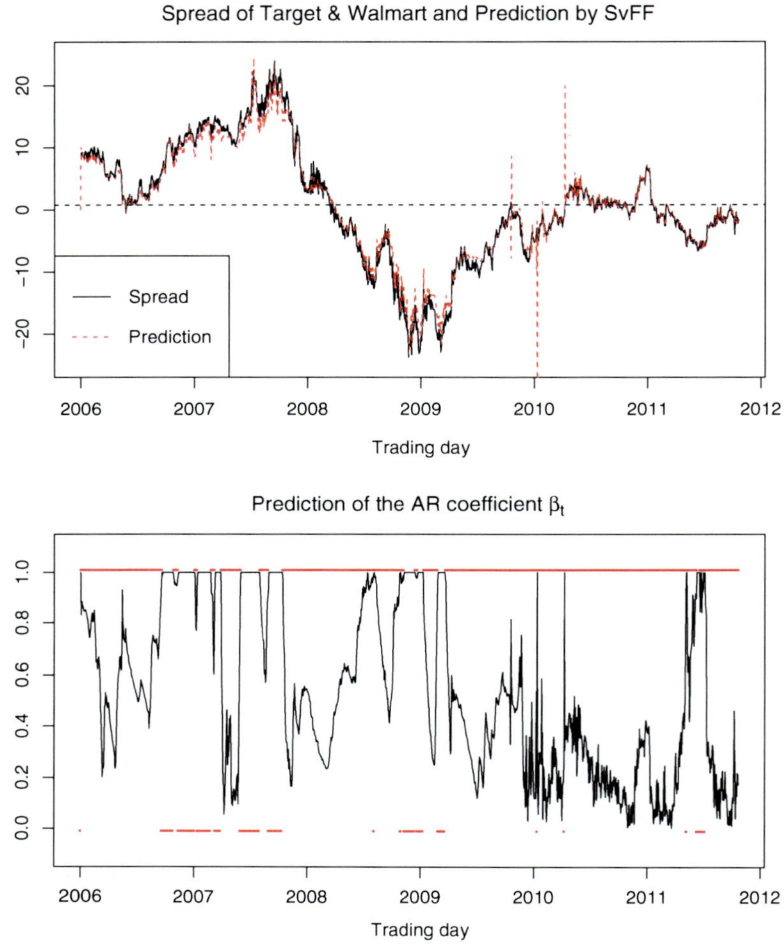

Fig. 5 Spread prediction using the SvFF algorithm (*upper panel*) and modulus of the predicted coefficients $\{\hat{\beta}_t\}$ using the SvFF algorithm (*lower panel*). Mean-reversion segments are detected for t satisfying $|\hat{\beta}_t| < 1$ and these regimes are highlighted with the *upper straight line* (the *lower straight line* indicates periods of breakdown of mean reversion)

produced by each algorithm. This table indicates that for any value $h = 1,3,5\,\%$ of $\hat{y}_{t+1}$, the BBvFF algorithm returns the highest cumulative profit. As h increases, the profits have the tendency to slightly reduce over the returns of each algorithm. This is due to the increased uncertainty in the prediction interval due to an increase in h. This is clearly depicted by the second part of the table, which again shows that as h increases the standard deviation of the cumulative profits increases. Now, for a given value of h, we observe that the standard deviation is smaller under the BBvFF algorithm. The table also shows the final balance; although for $h = 3,5\,\%$

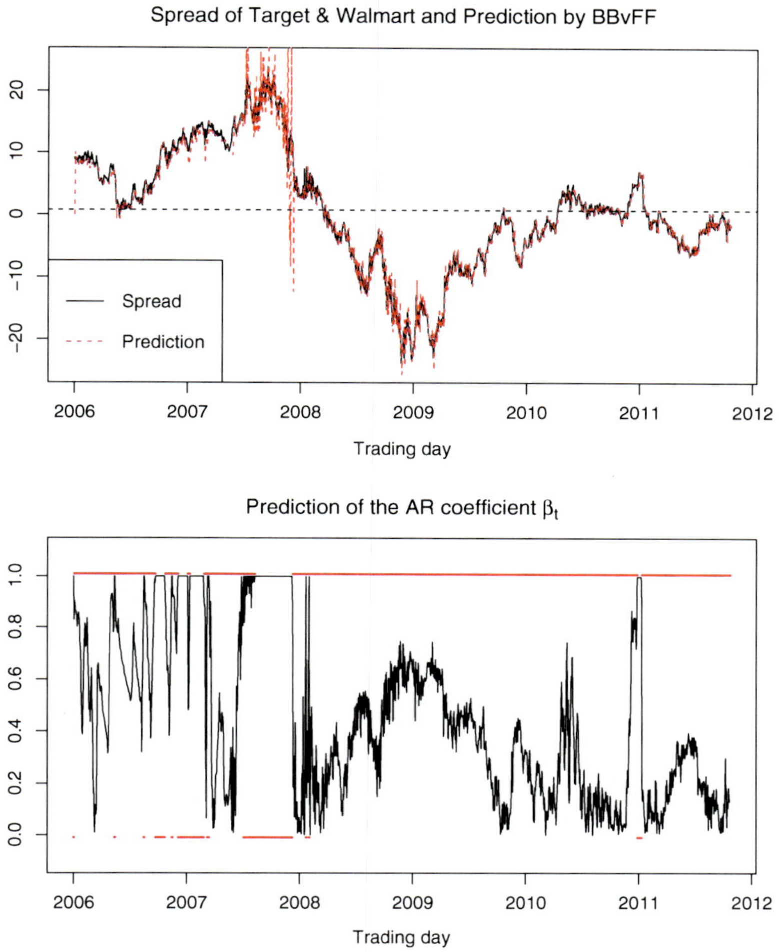

Fig. 6 Spread prediction using the BBvFF algorithm (*upper panel*) and modulus of the predicted coefficients $\{\hat{\beta}_t\}$ using the BBvFF algorithm (*lower panel*). Mean-reversion segments are detected for t satisfying $|\hat{\beta}_t| < 1$ and these regimes are highlighted with the *upper straight line* (the *lower straight line* indicates periods of breakdown of mean reversion)

the SDvFF algorithm shows the largest cumulative profit, one needs to put this result into perspective. Over time SDvFF results in loses as well as profits and as a result the mean cumulative profit is relatively low, in comparison with the other two algorithms. Figure 8 shows the cumulative profit over time for $h = 1\,\%$, produced by each of the three algorithms. We see that since May 2010, the BBvFF profits achieve a lower bound of around \$40, while for the same time period both SDvFF and SvFF reach the negative profit territory with the SDvFF reaching a loss of \$50, in the

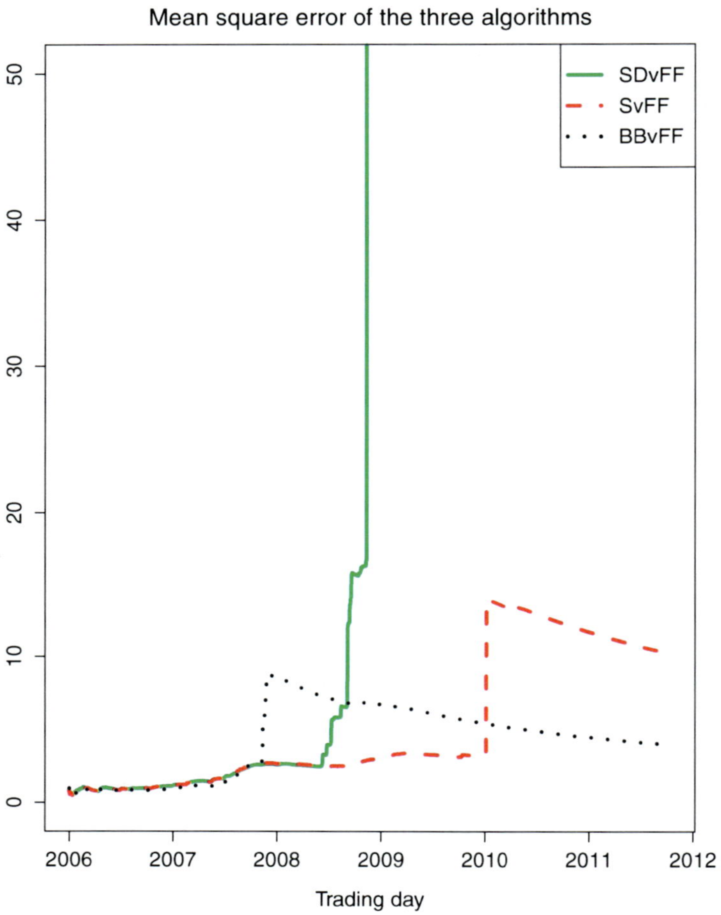

Fig. 7 Mean square error (MSE) computed over time from each of the three algorithms (SDvFF, SvFF and BBvFF)

beginning of 2011. Overall, we conclude that BBvFF with the proposed trading strategy offers the best performance, returning profit in excess of 127 % (minus transaction costs) and 130 % in mean cumulative profit.

6 Conclusions

In this paper we propose two new recursive least squares (RLS) algorithms with adaptive forgetting for the segmentation of mean-reversion in algorithmic pairs trading. The first algorithm is a combination of the usual RLS and Gauss–Newton

Table 1 Mean, standard deviation (STD) and final balance (FB) of cumulative profit, using each algorithm, for 3 values of the prediction margin h

	h		
Mean	1%	3%	5%
SDvFF	78.30	70.76	73.75
SvFF	92.33	77.04	66.53
BBvFF	130.92	112.54	115.21
STD	1 %	3 %	5 %
SDvFF	39.39	57.71	61.15
SvFF	32.03	46.03	52.89
BBvFF	32.19	38.80	44.69
FB	1 %	3 %	5 %
SDvFF	117.77	127.68	143.12
SvFF	123.25	117.62	109.85
BBvFF	127.80	119.98	133.71

RLS (SvFF), which switches from one algorithm to other according to the performance of the prediction errors. The second algorithm is an RLS combined with a new approach for the variable forgetting factor (VFF), according to which a binary process is used to classify low and high prediction error, and then adopting a conjugate Bayesian modelling approach based on beta and Bernoulli distributions. This approach appears to be more flexible than traditional VFF schemes as it provides the distribution of the VFF. A simple buy low and sell high trading strategy, based on the above algorithms, is described. The methods are illustrated by considering Target Corporation and Walmart Stores Inc share prices and the empirical findings suggest profits in excess of 130 % (minus transaction costs) over a time horizon of nearly 2 years.

The methodology of this paper may be suitable for other trading approaches, such as spread trading, in which the trader is interested in opening a trading position by speculating on whether the bid-ask spread of a single share will go up or down, for more information of which the reader is referred to [10].

Appendix

In this appendix we prove equations (5), (6) and (10).

We start with (5). From the definition of $\eta_t = \partial \psi_t / \partial \lambda$ and $L_t = \partial S_t / \partial \lambda$, we have

$$\begin{aligned}\eta_t &= \frac{\partial}{\partial \lambda}\{(I - P_t F_t F_t^{\mathrm{T}})\Phi\psi_{t-1} + S_t F_t e_t\} = -S_t F_t F_t^{\mathrm{T}}\Phi\psi_{t-1} + (I - P_t F_t F_t^{\mathrm{T}})\Phi\eta_{t-1} \\ &\quad + L_t F_t e_t + S_t F_t(-F_t^{\mathrm{T}}\Phi\psi_t) \\ &= (I - K_t F_t^{\mathrm{T}})\Phi\eta_{t-1} + L_t F_t e_t - 2S_t F_t F_t^{\mathrm{T}}\Phi\psi_{t-1},\end{aligned}$$

where $K_t = P_t F_t$ is used.

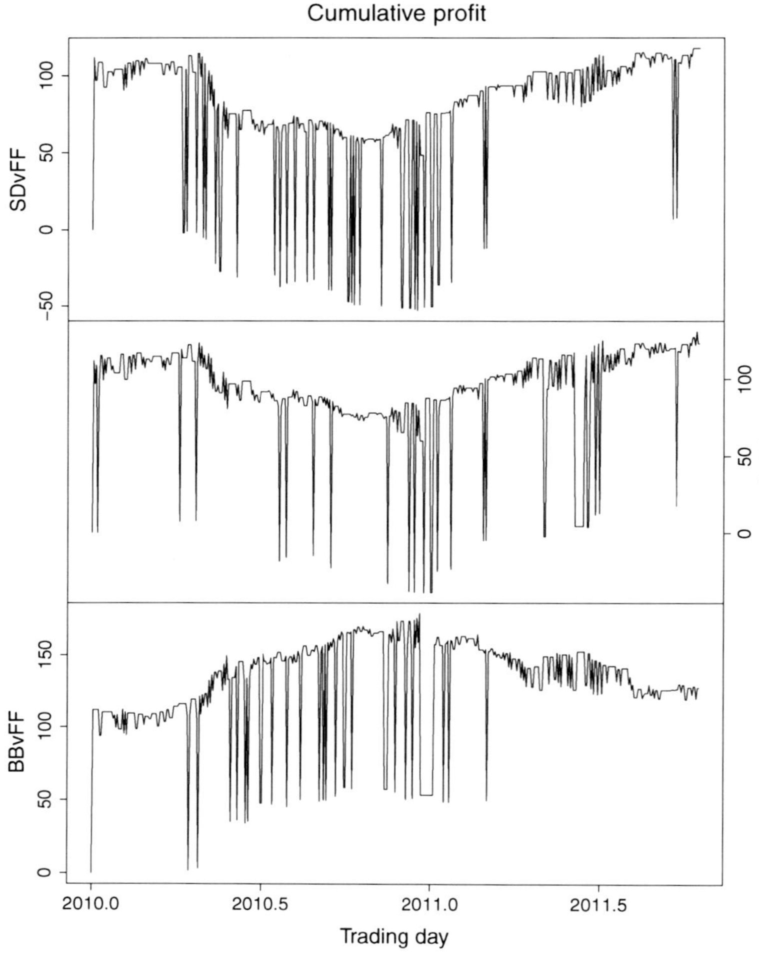

Fig. 8 Trading strategy implemented for each of the three algorithms (*top panel* for SDvFF, *middle panel* for SvFF and *lower panel* for BBvFF). Shown are the cumulative daily profits using $h = 1\,\%$

Next we prove (6).

$$\begin{aligned} L_t &= \frac{\partial}{\partial\lambda}\{-\lambda^{-1}P_t + \lambda^{-1}(I - P_tF_tF_t^{\mathrm{T}})\Phi S_{t-1}\Phi^{\mathrm{T}}(I - F_tF_t^TP_t) + \lambda^{-1}P_tF_tF_t^TP_t\} \\ &= \lambda^{-2}(I - K_tF_t^{\mathrm{T}})P_t - \lambda^{-1}S_t - \lambda^{-2}(I - K_tF_t^{\mathrm{T}})\Phi S_{t-1}\Phi^{\mathrm{T}}(I - F_tK_t^{\mathrm{T}}) \\ &\quad -\lambda^{-1}S_tF_tF_t^{\mathrm{T}}\Phi S_{t-1}\Phi^{\mathrm{T}}(I - F_tK_t^{\mathrm{T}}) + \lambda^{-1}(I - K_tF_t^{\mathrm{T}})\Phi L_{t-1}\Phi^{\mathrm{T}}(I - F_tK_t^{\mathrm{T}}) \\ &\quad -\lambda^{-1}(I - K_tF_t^{\mathrm{T}})\Phi S_{t-1}\Phi^{\mathrm{T}}F_tF_t^TS_t + \lambda^{-1}S_tF_tF_t^TP_t + \lambda^{-1}P_tF_tF_t^TS_t, \qquad (13) \end{aligned}$$

where $P_t - P_t F_t F_t^T P_t = (I - K_t F_t^{\mathrm{T}}) P_t$ and $K_t = P_t F_t$ are used. The proof of (6) is completed, by substituting in (13) M_t from (7).

Finally, we prove (10). For simplicity and convenience, we drop the time index in the distribution (9). To find the mode of λ, we find the maximum of $\log p(\lambda) = \log c + (c_1 - 1)\log(\lambda - \lambda_-) + (c_2 - 1)\log(\lambda_+ - \lambda)$. The first derivative with respect to λ is equal to

$$\frac{\partial \log p(\lambda)}{\partial \lambda} = \frac{c_1 - 1}{\lambda - \lambda_-} - \frac{c_2 - 1}{\lambda_+ - \lambda},$$

which by equalizing the above to zero returns the stationary point given by (10). The second partial derivative is always negative, i.e.

$$\frac{\partial^2 \log p(\lambda)}{\partial \lambda^2} = -\frac{c_1 - 1}{(\lambda - \lambda_-)^2} - \frac{c_2 - 1}{(\lambda_+ - \lambda)^2} < 0$$

and so the stationary point of (10) is a maximum.

Acknowledgements The paper has benefitted from discussions with Jeremy Oakley.

References

1. Cowles, A., Jones, H.E.: Some a posteriori probabilities in stock market action. Econometrica **5**, 280–294 (1937)
2. Elliott, R., Van Der Hoek, J., Malcolm, W.: Pairs trading. Quant. Finance **5**, 271–276 (2005)
3. Gatev, E., Goetzmann, W.N., Rouwenhorst, K.G.: Pairs trading: Performance of a relative-value arbitrage rule. Rev. Financ. Stud. **19**, 797–827 (2006)
4. Haykin, S.: Adaptive Filter Theory, 4th edn. Prentice Hall, Englewood Cliffs (2001)
5. Lowenstein, R.: When Genius Failed: The Rise and Fall of Long-Term Capital Management. Random House, New York (2000)
6. Malik, M.B.: State-space recursive least-squares with adaptive memory. Signal Process. **86**, 1365–1374 (2006)
7. Montana, G., Parrella, F.: In: Learning to Trade with Incremental Support Vector Regression Experts. Lecture Notes in Artificial Intelligence, vol. 5721, pp. 591–598. Springer, Berlin (2008)
8. Montana, G., Triantafyllopoulos, K., Tsagaris, T.: Flexible least squares for temporal data mining and statistical arbitrage. Expert Syst. Appl. **36**, 2819–2830 (2009)
9. Pole, A.: Statistical Arbitrage. Algorithmic Trading Insights and Techniques. Wiley, New Jersey (2007)
10. Pryor, M.: The Financial Spread Betting Handbook. Harriman, Petersfield, Great Britain (2007)
11. Song, S., Lim, J.-S., Baek, S., Sung, K.-M.: Gauss-Newton variable forgetting factor recursive least squares for time varying parameter tracking. Electron. Lett. **36**, 988–990 (2000)
12. Triantafyllopoulos, K., Montana, G.: Dynamic modeling of mean-reverting spreads for statistical arbitrage. Comput. Manag. Sci. **8**, 23–49 (2011)
13. Vidyamurthy, G.: Pairs Trading. Wiley, New Jersey (2004)
14. Zhang, H., Zhang, Q.: Trading a mean reverting asset: buy low and sell high. Automatica, **44**, 1511–1518 (2008)

Segmenting Carotid in CT Using Geometric Potential Field Deformable Model

Si Yong Yeo, Xianghua Xie, Igor Sazonov, and Perumal Nithiarasu

Abstract We present a method for the reconstruction of vascular geometries from medical images. Image denoising is performed using vessel enhancing diffusion, which can smooth out image noise and enhance vessel structures. The Canny edge detection technique, which produces object edges with single pixel width, is used for accurate detection of the lumen boundaries. The image gradients are then used to compute the geometric potential field which gives a global representation of the geometric configuration. The deformable model uses a regional constraint to suppress calcified regions for accurate segmentation of the vessel geometries. The proposed framework shows high accuracy when applied to the segmentation of the carotid arteries from CT images.

1 Introduction

The human circulatory system consists of vessels that transport blood throughout the body, providing the tissues with oxygen and nutrients. It is known that vascular diseases such as stenosis and aneurysms are often associated with changes in blood flow patterns and the distribution of wall shear stress. Modelling and analysis of the hemodynamics in the human vascular system can improve our understanding of vascular disease and provide valuable insights which can help in the development of efficient treatment methods. In recent years, computational fluid dynamics (CFD) has been widely used for patient-specific modelling of blood flow in vascular

S.Y. Yeo • I. Sazonov • P. Nithiarasu
College of Engineering, Swansea University, UK
e-mail: p.nithiarasu@swansea.ac.uk; i.sazonov@swansea.ac.uk

X. Xie (✉)
College of Science, Swansea University, UK
e-mail: x.xie@swansea.ac.uk

P. Latorre Carmona et al. (eds.), *Mathematical Methodologies in Pattern Recognition and Machine Learning*, Springer Proceedings in Mathematics & Statistics 30,
DOI 10.1007/978-1-4614-5076-4_10,

structures [6, 26, 29, 30]. Despite the involvement of numerous groups working in this field, and rapid advancement in efficient computational methods, there have been limited applications of computational hemodynamics in clinical practice. This is largely due to the challenges involved in the design of an integrated framework which can efficiently and accurately automate the interdisciplinary computational modelling process, which includes image segmentation, mesh generation and CFD simulation.

One of the main challenges in the computational modelling of hemodynamics is the accurate reconstruction of the vascular geometry. Anatomically accurate geometric models of the vascular structures are essential for realistic flow simulations and analysis. The anatomical information used to reconstruct the geometric models are usually provided in the form of medical image datasets (scans) from imaging modalities such as computed tomography (CT) and magnetic resonance (MR) imaging. Manual reconstruction of the vasculature geometries can be tedious and time consuming. There is also an issue of variability between the geometries extracted manually by different individuals, and variability of geometries extracted by the same individual at different occasions. In order to allow computational flow modelling to be efficiently applied as a diagnostic or predictive tool, the amount of user intervention required in the process should be reasonably small. In particular, a considerable amount of user intervention is often required in the reconstruction of an accurate geometric model for the simulation of flow dynamics. Therefore a robust and efficient method that can be used to accurately segment the geometric structures from medical image datasets can be very useful and advantageous in the modelling process. Here, we propose a robust framework for the segmentation of vessel geometries using the GPF deformable model. The framework is then applied to efficiently segment the geometries of carotid arteries from CT images.

Although several techniques exist for the segmentation of vascular structures from medical images, it remains an intricate process due to factors such as image noise, partial volume effects, image artifacts, intensity inhomogeneity and changes in topology. In [20], the coordinate points for the centreline of the aortic arch were extracted from volume-rendered MR images. A cubic spline was then used to represent the aortic centerline, and cross-sectional grids were generated on normal planes at equidistant points along the curve. This generated a curved tube with circular cross section of uniform radius, which is not representative of the geometry of the aorta. In [31], the centerline and diameter information of the vessels were extracted from the image dataset, and the vascular model was reconstructed using non-uniform rational B-splines (NURBS). Such techniques may often smooth out geometric information that can be important to the computation of accurate flow dynamics, such as those at bifurcations.

The 3D models of the vascular structures in [32] were reconstructed by extracting the 2D contours of the vessels at each of the image slices of the MR image dataset, and then lofting through the contours to create the surface models of the vessels. The different vessels were then merged using boolean operations in solid modelling. The cross sections of a particular vessel may, however, intersect with cross sections of branching vessels, and the geometry at these positions has to be approximated.

Other authors such as [4, 15, 23, 33, 39] also reconstructed 3D surface models of the vessels from 2D contours extracted from image slices. This sometimes requires positioning and orienting the 2D contours according to the medial axis of the vessels, and curve and surface interpolation are used to approximate and reconstruct the surface models. However, the 2D contour extraction techniques used do not provide control over 3D smoothness, and 3D geometric properties from the image datasets are not considered.

A simple thresholding technique was used in [21] to extract the binary image of the vessels, and the vascular model was reconstructed using an isosurface algorithm. The thresholding technique, however, does not consider the spatial characteristics of the image and is sensitive to image noise and inhomogeneous intensity. In [25, 37], region growing algorithms were applied to segment the vascular structures from CT and MR angiography data. The region growing techniques are, in general, sensitive to noise, and can often lead to non-contiguous regions and over-segmentation. In addition, thin structures are often not extracted due to variations in image intensities. The watershed transform was used in [1, 11] to extract the geometry of the carotid. In this approach, the image is interpreted as a landscape or topographic surface, with the pixel intensity representing the elevation of the topographic surface. Consider water on the landscape flowing towards regions with local minima, the watersheds are the lines that partition these regions. In this way, the image is partitioned into homogeneous regions with the watersheds defining the boundaries of the regions. The watershed transform tends to be sensitive to noise and often produces over-segmentation. It is also difficult for the watershed technique to extract thin structures and weak object edges.

In [14, 17], a 3D dynamic surface model was used to delineate the boundary of carotid arteries. An initial triangulated model was placed within the interior of the carotid vessels, and an inflation force was applied to deform the model towards the vessel wall. In particular, the inflation force is applied only when the vertices of the model are within the lumen, i.e., at locations with image intensity below a user-specified threshold. An image-based force is further applied to the surface model to better localize the boundary. It may, however, be difficult to select an appropriate threshold value that delineates the vessel wall closely due to inhomogeneous image intensity. This approach is sensitive to noise, and manual editing is often required to move the vertices towards the vessel wall. In [27], a 2D discrete dynamic contour was first used to extract the vessel contours, a dynamic surface model was then inflated to reconstruct the surface model using the binary images of the extracted contours. This, however, does not consider the 3D geometric information from the image dataset. In [5, 7, 38], the surface models for each of the vessel branches of the carotid artery were reconstructed independently using a tubular deformable model. A surface merging algorithm is then required to reconstruct the surface model of the carotid bifurcation from the triangulated surfaces of the vessel branches. This particular approach requires the determination of the axis of each of the vessels, which can be done manually by selecting a reasonable amount of points from image slices to represent the curves of the structure. Due to the smoothing effect of this technique, regions of high curvature such as those at bifurcations or stenosis may not be modeled accurately. These explicit deformable models represent contours and surfaces parametrically, which requires the tracking of points on the curves and surfaces during deformation. It is therefore difficult for explicit deformable models to deal with topological variation and complex shapes.

Implicit deformable models have been applied in the segmentation of vascular structures in [2, 3, 10, 22, 28]. However, many of these techniques use an attraction force field which acts on contours or surfaces only when they are close to the object boundaries. As such, initial contours have to be placed close to the object boundaries, which can be tedious in complex geometries. A constant pressure term such as the one in [18] is often used to monotically expand or shrink the deformable model towards the image object boundaries, which can overwhelm weak object edges. In addition, the initial contours have to be placed either inside or outside object boundaries, which can be difficult for compact and narrow structures. Many of these techniques are also sensitive to image noise and have difficulties in extracting deep boundary concavities.

2 Proposed Method

In this section, a robust framework is proposed for the reconstruction of vascular geometries from medical images. The approach consists of image denoising using vessel enhancing diffusion [12, 19], optimal edge detection using the Canny edge filter [8], and robust segmentation of the vascular geometries using GPF deformable model [36].

2.1 Vessel Enhancing Diffusion Filtering

The formulation of the vessel enhancing diffusion filter [12, 19] is based on a smoothed version of the vesselness measure used in [13]. In this approach, an anisotropic diffusion filter with strength and direction determined by the vesselness measure is applied to enhance the geometric structures of the vessel. The vesselness measure is determined by analyzing the eigensystem of the Hessian matrix given as:

$$\mathbf{H} = \begin{bmatrix} I_{xx} & I_{xy} & I_{xz} \\ I_{yx} & I_{yy} & I_{yz} \\ I_{zx} & I_{zy} & I_{zz} \end{bmatrix} \tag{1}$$

which describes the geometric information at each point of a 3D image I based on the local intensity variations. Here, the derivatives of the image I are computed as convolution with derivatives of the Gaussian function, i.e. $I_x = I(\mathbf{x}) * \frac{\partial}{\partial x} G_\sigma(\mathbf{x})$, where G_σ denotes the Gaussian function with standard deviation σ. The principal curvatures and directions are given by the maximum and minimum eigenvalues and the corresponding eigenvectors. With the eigenvalues given such that $|\lambda_1| \leq |\lambda_2| \leq |\lambda_3|$, the vesselness measure is defined as: if $\lambda_2 \geq 0$ or $\lambda_3 \geq 0$, $V_\sigma(\lambda) = 0$; otherwise

$$V_\sigma(\lambda) = \left(1 - \mathrm{e}^{\frac{-R_A{}^2}{2\alpha^2}}\right) \cdot \mathrm{e}^{\frac{-R_B{}^2}{2\beta^2}} \cdot \left(1 - \mathrm{e}^{\frac{-S^2}{2\gamma^2}}\right) \cdot \mathrm{e}^{\frac{2c^2}{|\lambda_2|\lambda_3{}^2}} \tag{2}$$

with

$$R_A = \frac{|\lambda_2|}{|\lambda_3|} \tag{3}$$

$$R_B = \frac{|\lambda_1|}{\sqrt{|\lambda_2\lambda_3|}} \tag{4}$$

$$S = \sqrt{\lambda_1{}^2 + \lambda_2{}^2 + \lambda_3{}^2} \tag{5}$$

in which R_A and R_B can be used to differentiate tubular structures from blob-like and plate-like structures, while S is used to differentiate between foreground vessel structures and background noise. The parameters α, β and γ are weighting factors which control the sensitivity of the vesselness measure, and c is a small constant.

For a multiscale analysis, the vesselness function is computed for a range of scales, and the maximum response is selected using the following equation:

$$V = \max_{\sigma_{\min} \leq \sigma \leq \sigma_{\max}} V_\sigma(\lambda) \tag{6}$$

A diffusion tensor is then defined such that vessel diffusion takes place in the direction of the vessel, while diffusion perpendicular to the vessel direction is inhibited. The diffusion tensor can therefore be used to preserve vessel structures and is given as:

$$\mathbf{D} = \mathbf{Q}\lambda'\mathbf{Q}^{\mathrm{T}} \tag{7}$$

where $\mathbf{Q}$ is a matrix containing the eigenvectors of the Hessian matrix $\mathbf{H}$, and λ' is a diagonal matrix with elements given as:

$$\lambda_1{}' = 1 + (w-1) \cdot V^{\frac{1}{s}} \tag{8}$$

$$\lambda_2{}' = \lambda_3{}' = 1 + (\epsilon - 1) \cdot V^{\frac{1}{s}} \tag{9}$$

with w, ϵ and s as tuning parameters. The anisotropic diffusion is then defined as:

$$L_t = \nabla \cdot (\mathbf{D}\nabla L) \tag{10}$$

where $L(0)$ is set as the input image. Figure 1 demonstrates that the vessel enhancing diffusion filter can be applied to enhance the vessel structures and smooth out noise in the image. The algorithm for the vessel enhancing diffusion filter has been implemented using the Insight Toolkit [16].

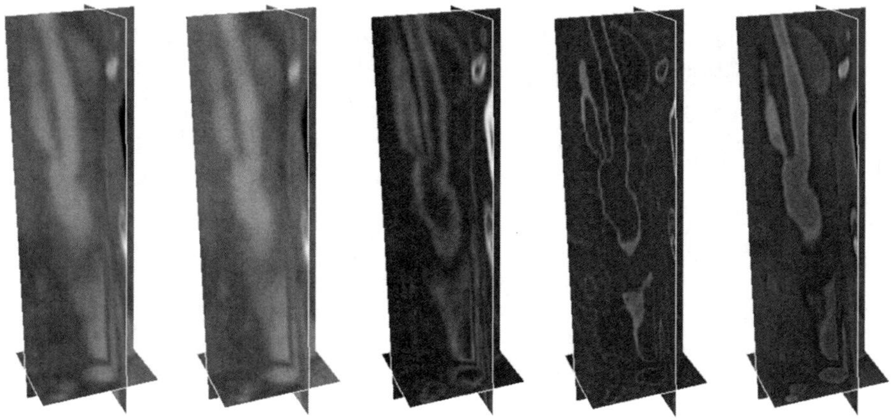

Fig. 1 Vessel enhancing diffusion and image object edge representation of CT image dataset 1, from left to right—original image, image with vessel enhancing diffusion, image gradient magnitude, Canny edge with image gradient intensities, geometric potential field

2.2 Edge Representation for Vessel Geometries

Image object edges are usually represented as regions with high intensity contrasts. Image gradients can be determined using the gradient operator or the Sobel filter. These techniques, however, produce object edge width of a few pixels. This can easily cause nearby structures to be connected. For complex geometries such as those in medical images, it is often necessary to determine fine edges using more robust edge detection techniques [9, 24] for accurate representation of the image structures. The Canny edge detection [8] can produce object edges with single pixel width, and can therefore be used for accurate edge detection of the vessel structures. In the Canny edge detection technique, image smoothing is first applied to reduce noise interference. This can be performed using the Gaussian filter or other smoothing techniques such as vessel enhancing diffusion [19]. The image gradients are then computed to determine the magnitudes and directions of the edges. Image pixels with magnitudes which are not local maxima in the directions of the edges are suppressed. Hysterisis thresholding is then applied to filter out spurious edges caused by noise. Image pixels with edge magnitude greater than a high threshold, i.e. $f_{\text{edge}}(x) > T_h$ are considered as edges, while pixels with edge magnitude lower than a low threshold, i.e. $f_{\text{edge}}(x) < T_l$ are removed. Image pixels with edge magnitudes in between the threshold values, i.e. $T_l \le f_{\text{edge}}(x) < T_h$, which are connected to edge pixels are also considered as edges. The image gradients at the detected edges are then used to compute the geometric potential field, see [36] for more detail. As shown in Fig. 1, the geometric potential field gives a more coherent representation of the image object boundaries as it utlizes global edge pixel interactions across the image.

2.3 Segmentation of Vessel Geometries using GPF Deformable Model with Region Constraint

It is shown in [34–36] that the GPF deformable model can be used to efficiently segment complex geometries from biomedical images. By using pixel or voxel interactions across the whole image domain, the deformable model is more robust to image noise and weak edges. The dynamic vector force field changes according to the relative postition and orientation between the geometries, which allows the deformable model to propagate through long tubular structures.

Here, the GPF deformable model is applied to segment the geometries of human carotid arteries from CT images. Some of the main challenges in the segmentation of the carotid geometries include intensity inhomgeneity, weak edges and adjacent veins with similar intensities to the carotids. In addition, calcifications which are attached to the arterial walls should not be included in the reconstructed geometries. Although the calcified plaques often appear as relatively bright regions compared to soft tissues, plaques with lower densities may have similar intensities to the lumen. As the intensities of the plaques vary with the densities, it is not easy for techniques such as global intensity threshold to remove the plaques from the extracted geometries. In this section, a region constraint is added to the deformable model such that it does not propagate across the calcified regions. This is done by constraining the deformable model from propagating across regions with image gradient magnitude larger than a user specified value, $T_{\max}$. As the calcified regions usually have relatively large image gradients, the threshold value can be easily selected by observing the histogram of the image gradient magnitude. The deformable model with region constraint can thus be expressed as:

$$\frac{\partial \phi}{\partial t} = \begin{cases} 0 & \text{if } |\nabla I| > T_{\max} \\ \alpha g \kappa |\nabla \phi| - (1-\alpha)(\mathbf{F} \cdot \nabla \phi) & \text{otherwise} \end{cases} \tag{11}$$

where α is a weighting parameter, g is the edge stopping function, κ is the curvature, and $\mathbf{F}$ is the geometric potential force defined in the GPF model [36].

Figures 2 and 3 depict a z-axis slice of the extracted geometry. As shown in the figures, some calcified regions have similar intensity to the lumen, which caused the deformable model to include them in the extracted geometries. The intensities of the plaques vary which makes it difficult for a global intensity threshold to suppress them. It is shown that by adding the region constraint, the deformable model can easily get around the calcified regions to segment the carotid geometries accurately.

3 Results and Discussion

In this section, experimental results on the segmentation of the cartoid geometries using the proposed framework are shown. In particular, six datasets from CT imaging (provided by Wolverhampton NHS trust) are used in the experiment.

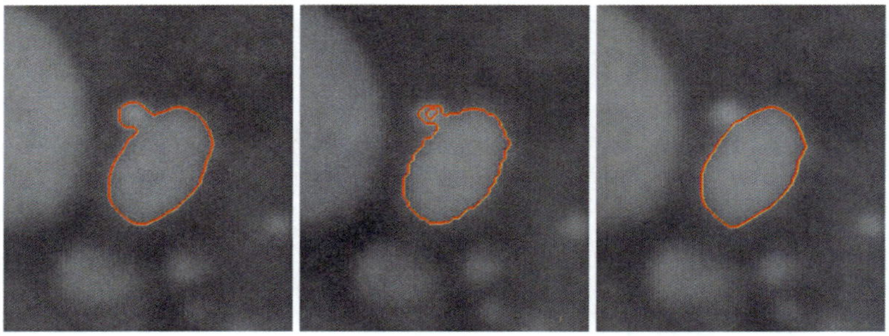

Fig. 2 Image slice from CT image dataset 2 showing contours (*top row*) and corresponding pixels (*bottom row*) extracted using: from left to right—GPF deformable model, GPF deformable model with intensity threshold and GPF deformable model with region constraint

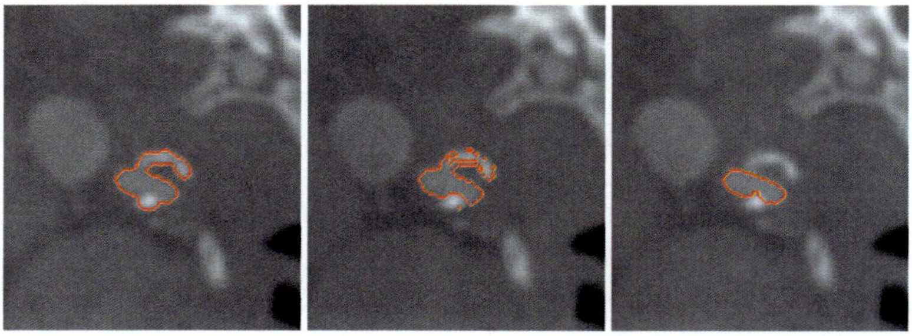

Fig. 3 Image slice from CT image dataset 4 showing contours (*top row*) and corresponding pixels (*bottom row*) extracted using: from left to right—GPF deformable model, GPF deformable model with intensity threshold and GPF deformable model with region constraint

The volumes of interest containing the carotid arteries are extracted from the image datasets to reduce the size of the input datasets. The robust framework which consists of vessel diffusion enhancing, computation of optimal object edge representation and deformable model with regional constraint is then applied for the reconstruction of vessel geometries.

Figures 4–7 depict the segmentation of the carotid geometries using the GPF deformable model with region constraint. As shown in Figs. 4 and 6, the bidirectional and dynamic vector force allows the flexible cross-boundary intializations of the model to easily propagate and converge to the geometries of the carotid arteries. The extraction of the vessel geometries from image datasets 1 and 4 took only 276 s and 494 s, while the extraction from image datasets 2 and 5 took 1,216 s and 1,379 s due to factors such as intensity variation, low constrast, multiple branches and complex topologies. A graphical user interface has been developed, which can be used to set multiple initial contours for fast convergence. It can also be used to remove inconsistency in object boundaries due to low resolution of the

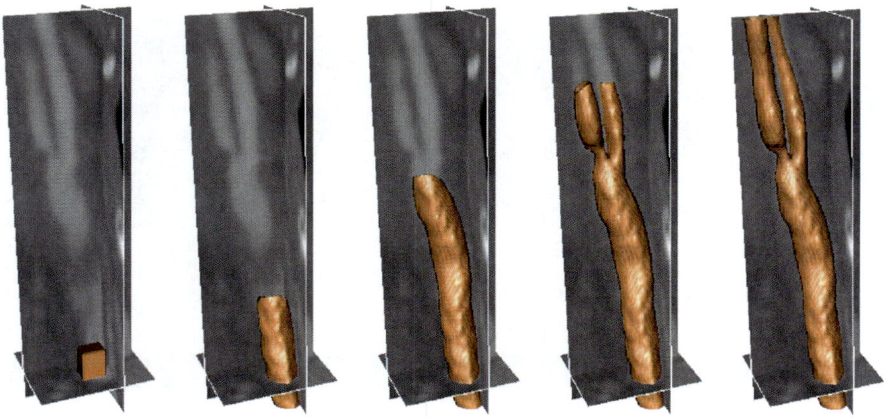

Fig. 4 Segmentation of carotid artery from CT image dataset 1 (61 × 71 × 125) using GPF deformable model (CPU-time, 276 s)

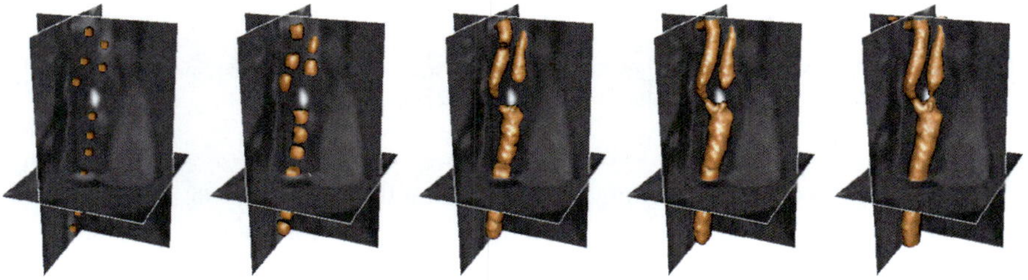

Fig. 5 Segmentation of carotid artery from CT image dataset 3 (70 × 80 × 120) using GPF deformable model (CPU-time, 206 s)

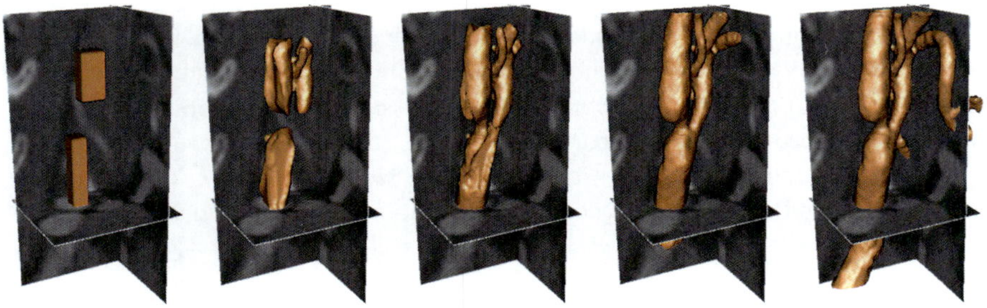

Fig. 6 Segmentation of carotid artery from CT image dataset 5 (70 × 80 × 120) using GPF deformable model (CPU-time, 1,379 s)

images, artifacts, etc., or small branches which do not affect the computational flow analysis. As shown in Figs. 5 and 7, one can easily speed up the segmentation process by placing multiple initial contours or surfaces, as the model converges to

Fig. 7 Segmentation of carotid artery from CT image dataset 6 ($70 \times 80 \times 120$) using GPF deformable model (CPU-time, 185 s)

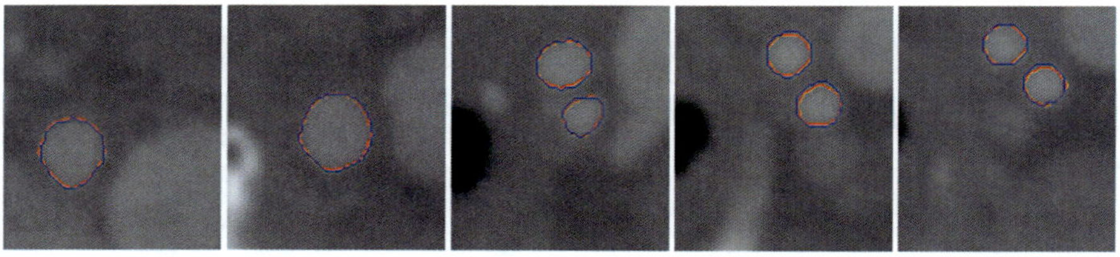

Fig. 8 Comparison of geometry segmented from CT image dataset 1 using image slices taken along z-axis direction: *blue*—manual, *orange*—GPF deformable model

the vessel geometries in 206 s and 185 s when applied to image datasets 3 and 6, respectively. Note that the deformable model easily propagates through the stenotic carotid bifurcations and get around the calcified regions to efficiently segment the carotid geometries from the CT images.

The reconstructed vessel geometries using the proposed framework are compared against geometries from manual segmentation. Figures 8–11 depict the comparison of the extracted geometries using random cross section slices taken along the z-axis direction. The blue and orange contours represent the cross section of the geometries extracted manually and using the GPF deformable model, respectively. As shown in the figures, the image dataset consists of other tissue structures which may affect the geometric reconstruction. In particular, vessels adjacent to the carotid artery can often cause other models to leak out due to the similar intensity. The geometric potential field provides a more coherent and global representation of the object edges and allows the deformable model to extract the geometry accurately. By adding a region constraint, the proposed model can easily get around the calcified regions as the deformable model propagates through the tubular structures to segment the vessel geometry as depicted in Figs. 9–11. The proposed framework can therefore be applied to segment the vessel geometries efficiently from the images. As shown in the figures, the vessel geometries segmented using the GPF deformable model with region constraint exhibit considerably small deviations from the manually extracted geometries.

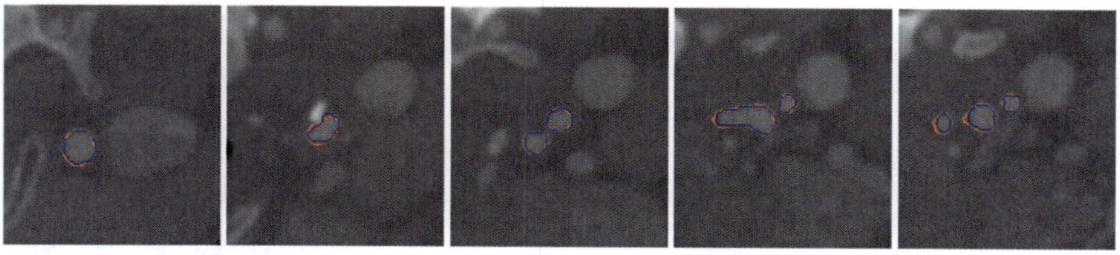

Fig. 9 Comparison of geometry segmented from CT image dataset 3 using image slices taken along z-axis direction: *blue*—manual, *orange*—GPF deformable model

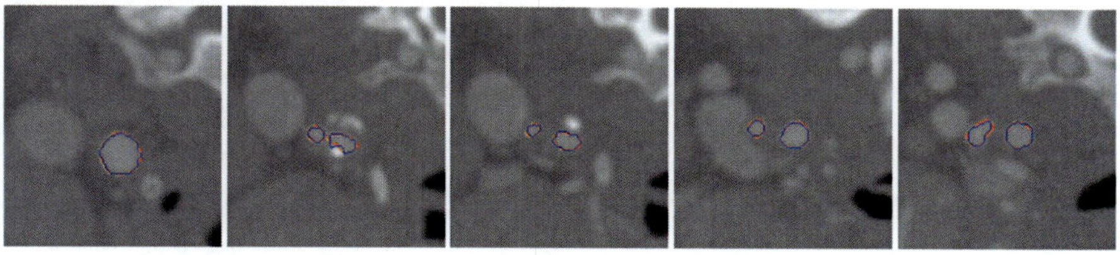

Fig. 10 Comparison of geometry segmented from CT image dataset 4 using image slices taken along z-axis direction: *blue*—manual, *orange*—GPF deformable model

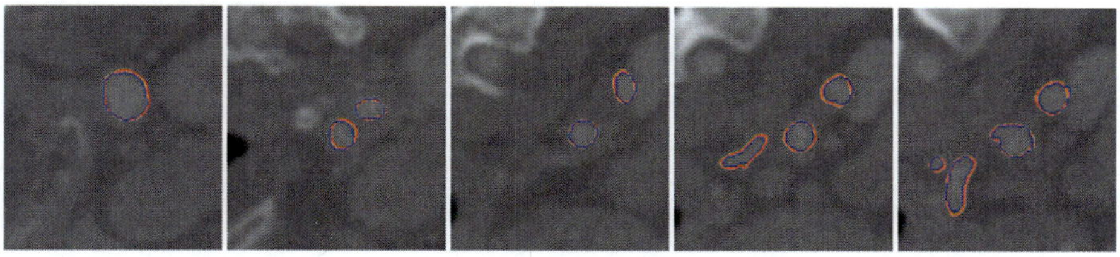

Fig. 11 Comparison of geometry segmented from CT image dataset 6 using image slices taken along z-axis direction: *blue*—manual, *orange*—GPF deformable model

Table 1 presents the accuracy of the segmented geometries using the proposed method. The foreground (FG) and background (BG) accuracy of the geometries was measured as the percentages of true foreground and background voxels which were segmented as foreground and background, respectively. The normalized overall accuracy is given as the average of FG and BG to measure the accuracy of correctly extracted voxels to reduce measurement bias towards the large number of background voxels in the image. It is shown that the proposed framework provides significantly accurate geometries with overall acurracies of 94.9 %, 94.8 %, 97.9 %, 99.5 %, 96.7 % and 97.0 % for image datasets 1–6, and an average overall accuracy of 96.8 %.

Table 1 Comparison of the segmented carotid geometries using the GPF deformable model with manual segmentation: Foreground (FG), background (BG) and overall accuracy measured in %

CT image dataset	GPF	
	FG (%)	89.9
1	BG (%)	99.9
	Overall (%)	**94.9**
	FG (%)	89.8
2	BG (%)	99.9
	Overall (%)	**94.8**
	FG (%)	96.0
3	BG (%)	99.9
	Overall (%)	**97.9**
	FG (%)	99.1
4	BG (%)	99.8
	Overall (%)	**99.5**
	FG (%)	93.8
5	BG (%)	99.5
	Overall (%)	**96.7**
	FG (%)	94.4
6	BG (%)	99.6
	Overall (%)	**97.0**
	FG average (%)	**93.9**
	BG average (%)	**99.8**
	Overall average (%)	**96.8**

References

1. Abdel-Dayem, A., El-Sakka, M.: Carotid artery ultrasound image segmentation using fuzzy region growing. In: International Conference on Image Analysis and Recognition, Springer, pp. 869–878 (2005)
2. Antiga, L., Ene-Iordache, B. Remuzzi, A.: Computational geometry for patient-specific reconstruction and meshing of blood vessels from mr and ct angiography. IEEE T-MI **22**(5), 674–684 (2003)
3. Antiga, L., Piccinelli, M., Botti, L., Ene-Iordache, B., Remuzzi, A., Steinman, D.A.: An image-based modeling framework for patient-specific computational hemodynamics. Med. Biol. Eng. Comput. **46**(11), 1097–1112 (2008)
4. Augst, A.D., Barratt, D.C., Hughes, A.D., McG Thom, S.A., Xy, X.Y.: Various issues relating to computational fluid dynamics simulations of carotid bifurcation flow based on models reconstructed from three-dimensional ultrasound images. Proc. Inst. Mech. Eng. H, J. Eng. Med. **217**(5), 393–403 (2003)
5. Cebral, J.R., Lohner, R., Soto, O., Choyke, P.L., Yim, P.J.: Patient-specific simulation of carotid artery stenting using computational fluid dynamics. In: MICCAI, Springer pp. 153–160 (2001)
6. Cebral, J.R., Hernandez, M., Frangi, A.F.: Computational analysis of blood flow dynamics in cerebral aneurysms from cta and 3d rotational angiography image data. In: International Congress on Computational Bioengineering, pp. 191–198 (2003)
7. Cebral, J.R., Castro, M.A., Lohner, R., Burgess, J.E., Pergolizzi, R., Putman, C.M.: Recent developments in patient-specific image-based modeling of hemodynamics. In: ENIEF04 (2004)
8. Canny, J.: A computational approach to edge detection. IEEE T-PAMI **8**(6), 679–698 (1986)
9. Deriche, R.: Using canny's criteria to derive a recursively implemented optimal edge detector. IJCV **1**(2), 167–187 (1987)

10. Deschamps, T., Schwartz, P., Trebotich, D., Colella, P., Saloner, D., Malladi, R.: Vessel segmentation and blood flow simulation using level-sets and embedded boundary methods. In: Computer Assisted Radiology and Surgery, pp. 75–80 (2004)
11. Ding, S., Tu, J., Cheung, C., Beare, R., Phan, T., Reutens, D., Thien, F.: Geometric model generation for CFD simulation of blood and air flows. In: International Conference on Bioinformatics and Biomedical Engineering, Elsevier pp. 1335–1338 (2007)
12. Enquobahrie, A., Ibanez, L., Bullitt, E., Aylward, S.: Vessel enhancing diffusion filter. Insight J., IEEE (2007)
13. Frangi, A.F., Niessen, W.J., Vincken, K.L., Viergever, M.A.: Multiscale vessel enhancement filtering. In: MICCAI, pp. 130–137, Springer (1998)
14. Gil, J.D., Ladak, H.M., Steinman, D.A., Frenster, A.: Accuracy and variability assessment of a semiautomatic technique for segmentation of the carotid arteries from three-dimensional ultrasound images. Med. Phys. **27**(6), 1333–1342 (2000)
15. Giordana, S., Sherwin, S.J., Peiro, J., Doorly, D.J., Papaharilaou, Y., Caro, C.G., Watkins, N., Cheshire, N., Jackson, M., Bicknall, C., Zervas, V.: Automated classification of peripheral distal by-pass geometries reconstructed from medical data. J. Biomech. **38**(1), 47–62 (2005)
16. Ibanez, L., Schroeder, W., Ng, L., Cates, J.: The ITK Software Guide, 2nd edn. Kitware, Inc. (2005)
17. Ladak, H.M., Milner, J.S., Steinman, D.A.: Rapid three-dimensional segmentation of the carotid bifurcation from serial MR images. J. Biomech. Eng. **122**(1), 96–99 (2000)
18. Malladi, R., Sethian, J.A., Vemuri, B.C.: Shape modelling with front propagation: A level set approach. IEEE T-PAMI **17**(2), 158–175 (1995)
19. Manniesing, R., Viergever, M.A., Niessen, W.J.: Vessel enhancing diffusion: A scale space representation of vessel structures. Med. Image Anal. **10**(6), 815–825 (2006)
20. Mori, D., Yamaguchi, T.: Construction of the CFD model of the aortic arch based on mr images and simulation of the blood flow. In: International Workshop on Medical Imaging and Augmented Reality, pp. 111–116 (2001)
21. Nanduri, J.R., Pino-Romainville, F.A., Celik, I.: CFD mesh generation for biological flows: Geometry reconstruction using diagnostic images. Comput. Fluids **38**(5), 1026–1032 (2009)
22. Nilsson, B., Heyden, A.: A fast algorithm for level set-like active contours. Pattern Recogn. Lett. **24**(9), 1311–1337 (2003)
23. Peiro, J., Sherwin, S.J., Giordana, S.: Automatic reconstruction of a patient-specific high-order surface representation and its application to mesh generation for CFD calculations. Med. Biol. Eng. Comput. **46**(11), 1069–1083 (2008)
24. Petrou, M., Kittler, J.: Optimal edge detectors for ramp edges. IEEE T-PAMI **13**(5), 483–491 (1991)
25. Sekiguchi, H., Sugimoto, N., Eiho, S., Hanakawa, T., Urayama, S.: Blood vessel segmentation for head MRA using branch-based region growing. Syst. Comput. Jpn. **36**(5), 80–88 (2005)
26. Steinman, D.A.: Image-based computational fluid dynamics modeling in realistic arterial geometries. Ann. Biomed. Eng. **30**(4), 483–497 (2002)
27. Steinman, D.A., Thomas, J.B., Ladak, H.M., Milner, J.S., Rutt, B.K., Spence, J.D.: Reconstruction of carotid bifurcation hemodynamics and wall thickness using computational fluid dynamics and mri. Mag. Resonan. Med. **47**(1), 149–159 (2002)
28. Svensson, J., Gardhagen, R., Heiberg, E., Ebbers, T., Loyd, D., Lanne, T., Karlsson, M.: Feasibility of patient specific aortic blood flow CFD simulation. In: MICCAI, Springer pp. 257–263 (2006)
29. Taylor, C.A., Figueroa, C.A.: Patient-specific modeling of cardiovascular mechanics. Ann. Rev. Biomed. Eng. **11**, 109–134 (2009)
30. Taylor, C.A., Steinman, D.A.: Image-based modeling of blood flow and vessel wall dynamics: Applications, methods and future directions. Ann. Biomed. Eng. **38**(3), 1188–1203 (2010)
31. Tokuda, Y., Song, M.H., Ueda, Y., Usui, A., Toshiaki, A., Yoneyama, S., Maruyama, S.: Three-dimensional numerical simulation of blood flow in the aortic arch during cardiopulmonary bypass. Eur. J. Cardio-thoracic Surg. **33**(2), 164–167 (2008)

32. Wang, K.C., Dutton, R.W., Taylor, C.A.: Improving geometric model construction for blood flow modeling. IEEE Eng. Med. Biol. Mag. **18**(6), 33–39 (1999)
33. Xu, X.Y., Long, Q., Collins, M.W., Bourne, M., Griffith, T.M.: Reconstruction of blood flow patterns in human arteries. Proc. Inst. Mech. Eng. H, J. Eng. Med. **213**(5), 411–421 (1999)
34. Yeo, S.Y., Xie, X., Sazonov, I., Nithiarasu, P.: Geometric potential force for the deformable model. In: BMVC, BMVA (2009)
35. Yeo, S.Y., Xie, X., Sazonov, I., Nithiarasu, P.: Level set based automatic segmentation of human aorta. In: International Conference on Computational and Mathematical Biomedical Engineering, CMBE (2009)
36. Yeo, S.Y., Xie, X., Sazonov, I., Nithiarasu, P.: Geometrically induced force interaction for three-dimensional deformable models. IEEE T-IP **20**(5), 1373–1387 (2011)
37. Yi, J., Ra, J.B.: A locally adaptive region growing algorithm for vascular segmentation. Int. J. Imag. Syst. Technol. **13**(4), 208–214 (2003)
38. Yim, P.J., Cebral, J.J., Mullick, R., Marcos, H.B., Choyke, R.L.: Vessel surface reconstruction with a tubular deformable model. IEEE T-MI **20**(12), 1411–1421 (2001)
39. Younis, H.F., Kaazempur-Mofrad, M.R., Chan, R.C., Isasi, A.G., Hinton, D.P., Chau, A.H., Kim, L.A., Kamm, R.D.: Hemodynamics and wall mechanics in human carotid bifurcation and its consequences for atherogenesis: investigation of inter-individual variation. Biomech. Model. Mechanobiol. **3**(1), 17–32 (2004)

A Robust Deformable Model for 3D Segmentation of the Left Ventricle from Ultrasound Data

Carlos Santiago, Jorge S. Marques, and Jacinto C. Nascimento

Abstract This paper presents a novel bottom-up deformable-based model for the segmentation of the Left Ventricle (LV) in 3D ultrasound data. The methodology presented here is based on Probabilistic Data Association Filter (PDAF). The main steps that characterize the proposed approach can be summarized as follows. After a rough initialization given by the user, the following steps are performed: (1) low-level transition edge points are detected based on a prior model for the intensity of the LV, (2) middle-level features or patch formation is accomplished by linking the low-level information, (3) data interpretations are computed (hypothesis) based on the reliability (belonging or not to the LV boundary) of the previously obtained patches, (4) a confidence degree is assigned to each data interpretation and the model is updated taking into account all data interpretations.

Results testify the usefulness of the approach in both synthetic and real LV volumes data. The obtained LV segmentations are compared with expert's manual segmentations, yielding an average distance of 3 mm between them.

Keywords 3D Echocardiography • Left ventricle • Segmentation • Deformable models • Feature extraction • Robust estimation • PDAF

1 Introduction

Ultrasound imaging plays an important role in the analysis of the cardiac function since it allows a real-time observation of the heart structures. The segmentation and tracking of the left ventricular (LV) endocardial border is a major goal in this context, since it provides information for measuring the ejection fraction and

C. Santiago (✉) • J.S. Marques • J.C. Nascimento
Institute for Systems and Robotics, Instituto Superior Tecnico, Lisbon, Portugal
e-mail: carlos.santiago@ist.utl.pt; jsm@isr.ist.utl.pt; jan@isr.ist.utl.pt

P. Latorre Carmona et al. (eds.), *Mathematical Methodologies in Pattern Recognition and Machine Learning*, Springer Proceedings in Mathematics & Statistics 30,
DOI 10.1007/978-1-4614-5076-4_11,

for assessing the regional wall motion [34]. A fully automatic LV segmentation system has the potential to streamline the clinical workflow and reduce the inter-user variability of the LV segmentation.

In ultrasound data, the LV appearance is mainly characterized by a dark region, representing the blood pool inside the chamber, enclosed by the endocardium, myocardium, and epicardium, which are roughly depicted by a brighter region (see Fig. 1). In ultrasonic devices, there is a great variability of the gray value distribution and the spatial texture in each of the above-mentioned regions. This happens among different ultrasound sequences and within the same sequence. This is due to the several reasons: fast motion during systole phase, low signal-to-noise ratio, edge dropout (specially in the diastole phase), the presence of shadows produced by the dense muscles, the specific properties and settings of the ultrasound machine, and the anisotropy of the ultrasonic image formation [6].

Some features make the problem of LV segmentation difficult. The large variation of the LV appearance forced researchers to impose constraints on the LV segmentation process using shape and motion models. Shape models are used to constraint the mean shape of the LV, as well as the main modes of variation, based on a collection of manually annotated LV images. However, the characterization of all possible shape patterns and variations has proven to be a difficult task given the large variability of LV shapes due to the heart anatomy. Another difficulty lies in the presence of outliers, that is, features in the image that do not belong to the LV boundary that hamper the LV shape estimates.

To tackle the above-mentioned issues, we use a 3D deformable model that is capable of large shape deformation at representing the LV contour, thus dealing with large variation of the LV appearance. To face with the presence of outliers (second difficulty), we exploit the use of Probabilistic Data Association Filter (PDAF) which is rooted in the seminal work of Bar Shalom [2]. Two main underlying ideas of the proposed algorithm are as follows. First *middle-level features* are considered, called patches, which consist of edge points grouped such that they form continuous surface portions. Second, each patch is labeled as being valid/invalid. Since we do not know beforehand the reliability of the patches, all possible labeling sequences of valid/invalid patch labels are considered. Each patch sequence is called here as *patch interpretation*. Finally, a probability (*association probability*) is assigned to each patch interpretation. Thus, in the adopted strategy, all patches contribute to the evolution of the deformable model with different weights.

The paper is organized as follows: in Sect. 2 we review the foremost ideas in this field of research; Sect. 3 presents an overview of the proposed segmentation system; Sect. 4 describes the deformable model used; Sect. 5 addresses the feature extraction algorithm and the middle-level features' assemblage; and Sect. 6 presents the robust model estimation technique. Section 7 shows results of the segmentation system applied to synthetic data and to the segmentation of the LV in echocardiographic images. Finally, Sect. 8 concludes the paper with final remarks about the developed system and future research areas.

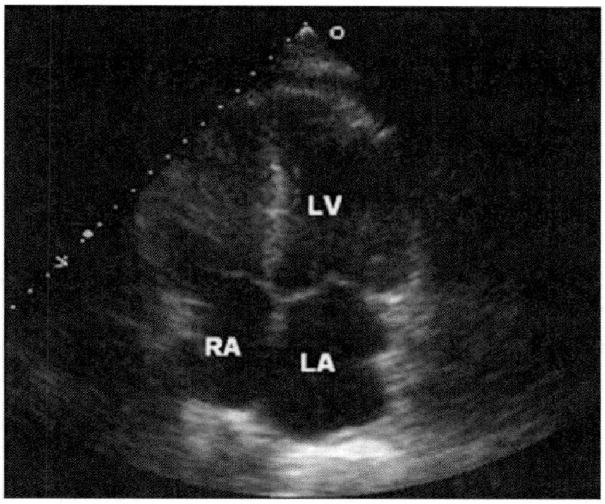

Fig. 1 Echocardiography—apical four-chamber view [10]

2 Previous Work

The literature concerning the segmentation of the LV is large. The main techniques that have been successfully addressed can be divided into the following classes: (1) bottom-up approaches [41, 45], (2) active contours methods [25], (3) active shape models (ASM) [12], (4) deformable templates [11, 17, 32, 44], (5) active appearance models (AAM) [7, 13, 30], (6) level set (LS) approaches [5, 14, 15, 27, 28, 36, 37, 39], and (7) database-guided (DB-guided) segmentation [8, 9, 19, 46].

Bottom-up approaches detect the LV boundary using edge detection that constitutes features to represent the object boundary. Although these methods have low computational complexity, they are sensitive to initial conditions and generally lack robustness to imaging conditions.

Active contours methods inspired the development of level set (LS) methods [29], which significantly reduce the sensitivity to initial conditions. The use of level sets for medical image segmentation aims at improving the performance of active contours due to the following. First, LS are able to increase robustness of the model by combining both region and boundary segmentation. Second, the texture and shape priors are jointly used with a continuous parametric function to model the implicit segmentation function [4, 5, 14, 27, 28, 37, 39].

Alternatively, these issues can also be faced using deformable templates [11, 17, 32, 44] that use an unsupervised scheme for learning. However, deformable template-based methods require the knowledge of how the initialization is performed.

Both level-sets and deformable templates have demonstrated good results when dealing with medical images. Nonetheless, they also present some drawbacks regarding the prior knowledge included in the optimization function.

The previously raised issues have also motivated the development of the supervised based models, in which the shape and appearance of the LV are learned from a manually annotated training set. This class of methods include the active shape model (ASM) [12] and active appearance model (AAM) [7, 13, 30]. However, both methods need a large set of annotated training images and the initialization must be close to a local optimum. Furthermore, these methods assume a Gaussian distribution of the shape and appearance derived from the training samples. The above issues motivated the proposal of the DB-guided segmentation approaches that use supervised learning techniques [19, 46]. Specifically, discriminative learning model based on boosting techniques [18] is developed to segment LV from ultrasound images. Another important point in the DB-guided approach is its independence regarding an initial guess. Instead of that, a full search is conducted in the parameter space. However, these methods have several shortcomings. Besides the high complexity of the search process, supervised learning methods face two main difficulties which are the large number of training images (in the order of hundreds) needed to estimate the parameters of the model and the robustness to imaging conditions absent from the training set.

Nonetheless, 3D echocardiography has gained increasing interest and several methods to perform the 3D segmentation of the LV have become available in literature [34]. One approach to perform 3D segmentation is to consecutively applying 2D segmentations to each image plane and assembling them into a 3D structure [33, 40]—as cardiologists manually do in such cases. However, such approaches require additional methods to prevent spacial inconsistencies in the surface. Other approaches have performed the 3D segmentation using 3D active contours such as level-set [21, 22, 24]. Furthermore, over the last decade some effort has been put into developing 3D+t LV tracking systems that are able to segment the LV over the course of the cardiac cycle, such as [35, 42].

3 System Overview

The idea behind the present approach is to tackle the difficulties of classic deformable contour methods associated with noisy images (such as ultrasound images) by introducing a robust estimation scheme. The robust framework is inspired in the S-PDAF [31], developed for shape tracking in cluttered environments. Here we extend it to the context of 3D shape estimation.

The proposed segmentation system uses a 3D deformable model to characterize the surface of the segmentation. This deformable surface requires an initialization procedure that ensures it is initialized in the vicinity of the LV boundary.

The adaptation procedure is an iterative process that consists of the following steps. After initialization of the model, an adaptation cycle begins with the detection of low-level features (edge points), searched in the vicinity of the model. Then, these are grouped into middle-level features (patches) to form continuous and meaningful surface portions. Based on the assembled patches, the S-PDAF algorithm determines all possible interpretations of considering a patch valid or invalid and assigns each patch interpretation a confidence degree. This confidence

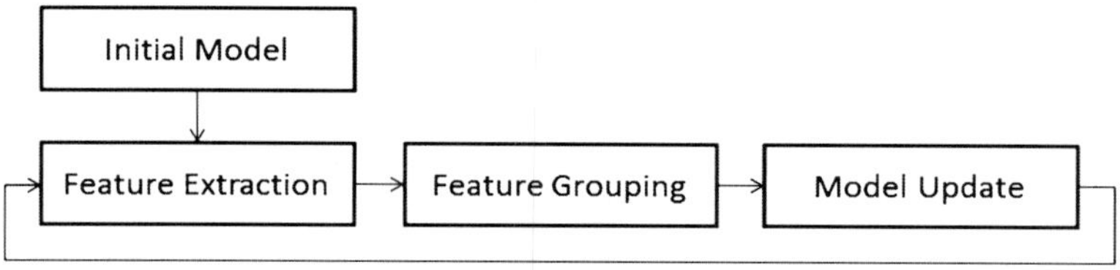

Fig. 2 Diagram of the proposed segmentation system

degree determines the weight of the patch interpretations in the estimation of the LV boundary location. The surface model then adapts towards the estimated LV boundary, ending an iteration of the adaptation cycle. The process repeats until the surface is considered close to the LV boundary. Figure 2 shows a diagram of the adaptation cycle.

4 Surface Model

The proposed segmentation system uses a 3D deformable model called simplex mesh [16]. A 3D simplex mesh is a meshed surface composed of vertices and edges, where each vertex has three neighbors (i.e., belongs to three different edges) (see Fig. 4). This particular structure allows the definition of geometric relations between vertices that are used in the adaptation procedure to ensure a smooth surface and good vertex distribution.

4.1 *Law of Motion*

Each vertex P_i adapts by an iterative process (with iteration number k) under the influence of external and internal forces, and its final position is determined by the equilibrium of these forces using the following equation [16]:

$$P_i(k+1) = P_i(k) + (1-\gamma)(P_i(k) - P_i(k-1)) + \alpha_i \mathbf{F}_i^{\text{int}}(k) + \beta_i \mathbf{F}_i^{\text{ext}}(k) \qquad (1)$$

where the parameters γ, α and β are constants.

The internal force, F^{int}, is responsible for maintaining the smoothness of the surface, making use of the geometric relations between vertices. On the other hand, the external force, F^{ext}, is responsible for attracting each vertex towards the object boundary.

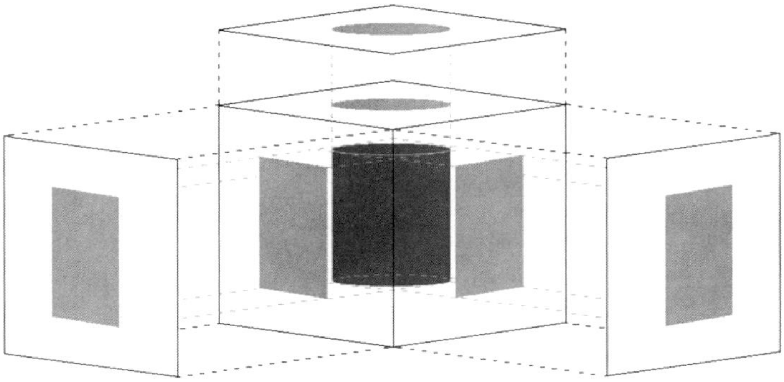

Fig. 3 Schematic example of the creation of a carved volume (*cylinder*). The *light gray* areas correspond to the binary mask created by the manual segmentation, whereas the *dark gray* volume corresponds to the intersection of the three carved volumes

4.2 Model Initialization

The initialization procedure has to meet the following conditions: (1) the initial model should be initialized in the vicinity of the LV boundary and (2) it should be a simplex mesh. These two conditions are met using the following three steps.

First, to ensure that the model is initialized in the vicinity of the LV boundary, the user manually defines a coarse outline of the LV in three orthogonal planes. A 3D region is then obtained by space carving [26] (see an example in Fig. 3).

Second, the simplex mesh is initialized as a sphere in the center of the carved volume. After uniformly sampling sphere points, the convex hull algorithm [1] is applied, resulting in a triangular mesh on the sphere surface. Then, taking into account the duality between simplex meshes and triangulations [16], an associated simplex mesh can be formed by considering the center of each triangle as vertices and linking each vertex with the center of the three neighboring triangles, resulting in the simplex mesh shown in Fig. 4a.

Finally we let the spherical simplex mesh deform until it fits the carved region. This region corresponds to the silhouette of the LV boundary and simplifies an initial adaptation to the dataset since it is a noiseless binary volume.

5 Feature Extraction

The detection of the LV border is performed by the feature extraction algorithm. First, each slice of the volume is pre-processed using a 2D median filter with a window size of 4×4 pixels (alternatively, a 3D median filter can also be used). Feature extraction is then performed using a directional feature search in the vicinity of the surface model.

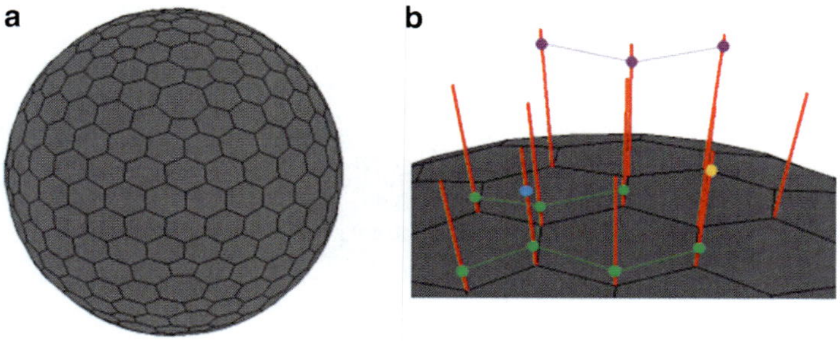

Fig. 4 (**a**) Simplex mesh initialized as a sphere. (**b**) Detail of simplex mesh showing the search lines (*red lines*) and detected features (*colored dots*); each dot color and colored links between them correspond to a different patch

5.1 *Feature Detection*

At each vertex, P_i, of the simplex mesh, we compute the normal to the surface and define a search line parallel to the normal vector that passes through P_i (See Fig. 4b). Then, the intensity signal along the search line is analyzed. The LV border features are detected using an edge detector filter, as described in [3]. The filter output's maxima are extracted using a threshold and a non-maximum suppression technique.

Although this methodology has good results, many undesired features are still detected, depending on the threshold used. In our experimental setup, we detect up to four features per search line, and recall that only one of these features corresponds to the LV boundary.

5.2 *Middle-Level Features*

To increase the robustness of the feature detection, these are grouped into middle-level features (patches). To assemble these patches, we assign to each feature a patch label. The labeling algorithm assumes that features should belong to the same patch if: (1) the corresponding vertices are neighbors (*i.e.*, the patch is a connected graph of its features); (2) all features associated with the same vertex (located along the same search line) have different labels; and (3) the distance between neighboring features in a patch must not exceed a chosen threshold. Figure 4b shows an example of the labeling result.

In order to achieve the desired label configuration, L, we define an energy function $E(L)$ composed of three terms. The first term, $E_1(L)$, is minimum when features with the same label are the closest features associated with the neighboring vertices. The second term, $E_2(L)$, prevents patches from having features too far apart. This is done by assigning an energy of ∞ to labels where the distance between neighboring features exceeds the labeling threshold. If the distance is lower than the

Table 1 Labeling algorithm

```
Q = {} % labeling queue
C = {} % labeled features
repeat
If Q is empty
    seed a new label l in a random feature y^i ∉ C
    add y^i to C
    for each feature y^j neighbor of y^i
        if y^j ∉ C & labeling y^j with l lowers E(L)
            add y^j to Q
Else
    repeat
    y^i = Q(1)
    label y^i with l
    add y^i to C
    remove y^i from Q
    for each feature y^j neighbor of y^i
        if y^j ∉ C & labeling y^j with l lowers E(L)
            add y^j to Q
    until Q is empty
until all features have been labeled
```

threshold, the energy yields the value 0. Finally, the third term, $E_3(L)$, prevents repeated labels in features associated with the same vertex, again assigning an energy value of ∞ if this occurs and 0 otherwise. The total energy of the label configuration is:

$$E(L) = E_1(L) + E_2(L) + E_3(L) \tag{2}$$

The label configuration L that minimizes the total energy function (2) corresponds to the configuration that obeys all the mentioned conditions.

To determine the label configuration that minimizes the total energy function, we resort to a labeling algorithm that uses a region growing scheme. In this algorithm, a new label is seeded in a random feature and it propagates to the surrounding features whenever an energy decrease is possible. This process repeats until all features have been labeled. The pseudocode in Table 1 describes the developed labeling algorithm.

The size of the resulting patches and their distance to the surface provides good differentiation measures to assess if the features in that patch belong to the LV boundary or if they were produced by the background.

6 Robust Model Estimation

The robust model estimation used is an extension of the S-PDAF algorithm described in [31] to the 3D case. The algorithm consists of the following. In each iteration k, this estimation technique considers all possible combinations of valid/invalid labels for the patchs. Assuming $M(k)$ patches were detected in iteration k, there are $m(k) = 2^{M(k)}$ possible interpretations. Each combination is defined as a patch interpretation $I_i(k) = \{I_i^1(k), \ldots, I_i^n(k), \ldots, I_i^{M(k)}(k)\}$, where $I_i^n(k) = 0$ if the nth patch in the ith interpretation is considered invalid and $I_i^n(k) = 1$ otherwise.

The model assumes that the LV boundary position is described by the state vector $x(k)$, which contains the 3D coordinates of all the simplex mesh's vertices, and follows the motion model:

$$x(k) = x(k-1) + w(k) \quad (3)$$

where $w(k) \sim N(0, Q)$ is a realization of a random variable with normal distribution (white noise).

For each interpretation $I_i(k)$, the observations $y_i(k)$ are generated by a specific model. If an observation $y_i(k)$ is considered invalid (outlier), the model assumes it is generated by uniform distribution. Otherwise, the model assumes it relates to the boundary points $x(k)$ by:

$$y_i(k) = C_i(k)x(k) + v_i(k) \quad (4)$$

where $C_i(k)$ is the observation matrix that relates the vertices to the corresponding observations, and $v_i(k) \sim N(0, R_i)$ is white noise with normal distribution associated with the valid features $y_i(k)$ of the interpretation $I_i(k)$.

The state estimate is then defined by [31]:

$$\hat{x}(k) = \sum_{i=0}^{m_k} \hat{x}^i(k)\alpha_i(k) \quad (5)$$

where $\hat{x}^i(k)$ is the updated state conditioned on the hypothesis that $I_i(k)$ is correct. This updated state is computed in the same way as the update state equation of a traditional Kalman filter. $\alpha_i(k)$ is the association probability of the interpretation $I_i(k)$. A similar analysis is done to update the covariance matrix [31].

The obtained estimate is used in the computation of the external force, $\mathbf{F}_i^{\text{ext}}(k)$, in (1). The desired position of a vertex $P_i(k)$ is the corresponding estimate $\hat{x}_i(k)$, therefore:

$$\mathbf{F}_i^{\text{ext}}(k) = \hat{x}_i(k) - P_i(k) \quad (6)$$

6.1 Association Probabilities

The association probabilities, $\alpha_i(k)$, define the strength of the corresponding interpretation $I_i(k)$ in each iteration k of the adaptation procedure. We assume that the association probabilities depend on the following variables: $L^k = \{L(k), L^{k-1}\}$, which is the set of all patches from iteration 1 until k, where $L(k) = \{L_1, \ldots, L_{M(k)}\}$

is the set of detected patches in iteration k; and $A(k) = \{A_1, \ldots, A_{M(k)}\}$ is the set of patch areas, which correspond to the number 0 (low-level) features each patch includes—$L_n = \{y_1^n, \ldots, y_{A_n}^n\}$. Then, we define the association probability of an interpretation $I_i(k)$ as:

$$\alpha_i(k) = P(I_i(k)|L(k), L^{k-1}, M(k), A(k)) \quad (7)$$

From this point on, we omit the dependence on (k) for the sake of simplicity. Using a Bayesian approach, this probability can be decomposed into:

$$\alpha_i = \frac{P(L|I_i, A, M, L^{k-1}) \times P(I_i|A, M, L^{k-1})}{\beta} \quad (8)$$

where $\beta = P(L|A, M, L^{k-1})$ is a normalization constant that does not depend on I_i, $P(L|I_i, A, M, L^{k-1})$ is the likelihood of the set of patches L and $P(I_i|A, M, L^{k-1})$ is the prior probability of the interpretation I_i conditioned on the patches (i.e., based on its valid and invalid patches). Furthermore, an association probability of 0 will be assigned to interpretations with overlapping (valid) patches and patches considerably smaller than the larger ones will be promptly discarded to avoid an exponential growth of possible interpretations.

Assuming that the patches are independently generated, the likehood can be expressed as $P(L|I_i, A, M, L^{k-1}) = \prod_n P(L_n|I_i, A, M, L^{k-1})$, where each individual term is the probability of having a patch L_n at an average distance d to the surface. This probability is dependent on the hypothesis that L_n is considered valid or invalid: if $I_i^n = 0$ it is assumed that the probability distribution is uniform along the search line, whereas if $I_i^n = 1$ the probability distribution is assumed Gaussian with mean 0 and covariance proportional to the length of the search line, V. Formally:

$$P(L_n|I_i, A, M, L^{k-1}) \sim \begin{cases} V^{-1} & \text{if } I_i^n = 0 \\ \rho^{-1} N(d; 0, \sigma) & \text{otherwise} \end{cases} \quad (9)$$

where ρ is the normalization constant.

As to the prior probability of each interpretation I_i, it is related to the area of its valid and invalid patches. The probability $P(I_i|A, M, L^{k-1})$ can also be decomposed as the product of each individual probability of the patches, $P(I_i^n|A, M, L^{k-1})$. It is assumed that larger patches are more likely to belong to the LV boundary. Therefore, these should receive a higher probability. On the other hand, when considered invalid, these should be assigned a small probability.

The resulting prior probability yields:

$$P(I_i|A, M, L^{k-1}) = \prod_{L_n : I_i^n = 1} [a \log(A_n + 1) + b] \times \prod_{L_n : I_i^n = 0} 1 - [a \log(A_n + 1) + b)] \quad (10)$$

where:

$$\begin{aligned} a &= \frac{P_A - P_B}{1 - \log(A_{\max}+1)} \\ b &= P_B - a\log(A_{\max}+1). \end{aligned} \tag{11}$$

and P_A, P_B, and $A_{\max}$ are constants. These assumptions assure that patches with large areas receive a high prior probability.

7 Results

The proposed system was tested using different datasets, both real and synthetic. The purpose of the synthetic data was to assess the functionality of the model. The real data consists of four echocardiographic volumes, with $208 \times 208 \times 224$ voxels and a resolution in x, y, and z of, respectively, 0.834 mm, 0.822 mm, and 0.726 mm. The algorithm was applied to four different echocardigraphic volumes.

A quantitative assessment of the system's performance will be provided using error metrics between the obtained segmentation and the manual segmentation performed by an expert—considered as the ground truth (GT). We use four similarity metrics to compare the output of the algorithm with the reference contours, namely: the Hammoude metric [20], d_{HMD}, the average distance, d_{AV}, the Hausdorff metric [23], d_{HDF}, and mean absolute distance, d_{MAD}. These are computed as described in [38].

7.1 *Parameter Definition*

The parameters of the model were fine-tuned and kept constant in all the tests. In (1), the constants are set to $\alpha = 0.7$, $\beta = 0.05$ and $\gamma = 0.9$; the adaptation process stops when the average displacement of the vertices went bellow 0.005 voxels. In the feature extraction algorithm, the threshold used in the maxima detection was $t_f = 0.5c_{\max}$, where $c_{\max}$ is the highest peak of the filter output; as to the labeling threshold (i.e., the maximum distance allowed between neighboring features with the same label) adopted was $t_l = 8$. Finally, in (10) we used $P_A = 0.05$, $P_B = 0.95$ and $A_{\max} = 700$ (the number of vertices in the surface).

7.2 *Synthetic Data*

We present one particular test using a synthetic volume. This synthetic volume contains a sphere corrupted by white Gaussian noise with zero mean (see Fig. 5). Although many features are detected (an average of approximately three features

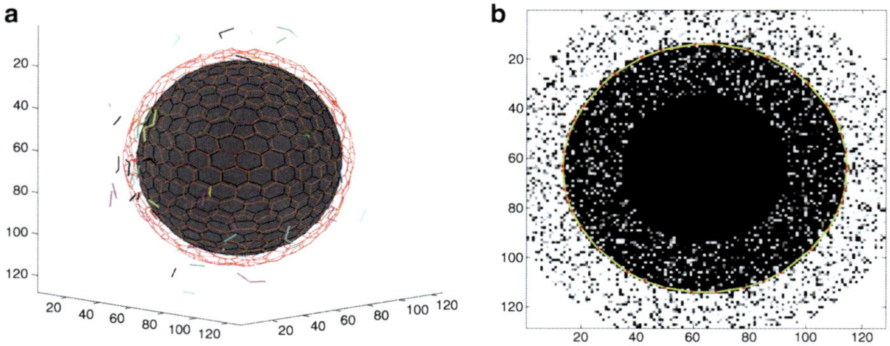

Fig. 5 Segmentation of synthetic data containing a noisy sphere. (**a**) Patches detected (*colored meshes*) in the vicinity of the surface model (*gray mesh*). (**b**) Slice view of the volume and corresponding cross section of the surface after the adaptation cycle (*yellow contour*)

per search line), only one large patch is extracted (see Fig. 5a) and all the other noise-originated patches are discarded due to their smaller size. Therefore, only two interpretations are possible: one that considers the large patch valid and the other that considers it invalid. The association probabilities of the existing interpretations are the following:

$$\alpha_1 = P(I_1 = \{I^1 = 0\}) = 0.04$$
$$\alpha_2 = P(I_2 = \{I^1 = 1\}) = 0.96$$

which means a high confidence degree is assigned to the large (correct) patch. Figure 5b shows that the model is able to correctly fit the desired sphere.

7.3 Echocardiographic Data

As mentioned before, the segmentations of the LV in four different echocardiographic volumes are presented in Fig. 6. For each volume, a single slice is shown twice: one containing the estimated contour (left), obtained by intersecting the simplex mesh with a plane, and the GT contour (center). The final three-dimensional surface is also presented.

Figure 6 shows that the developed segmentation system performs reasonably well. A quantitative evaluation of the results was performed using the similarity metrics mentioned above and the results can be seen in Table 2.

The table shows high similarity between the estimated contour and the GT, with an average distance ($\bar{d}_{\text{AV}}$) of 3 mm between the closest points and an average error ($\bar{d}_{\text{HMD}}$) of 17.5%. This indicates a good match between both contours.

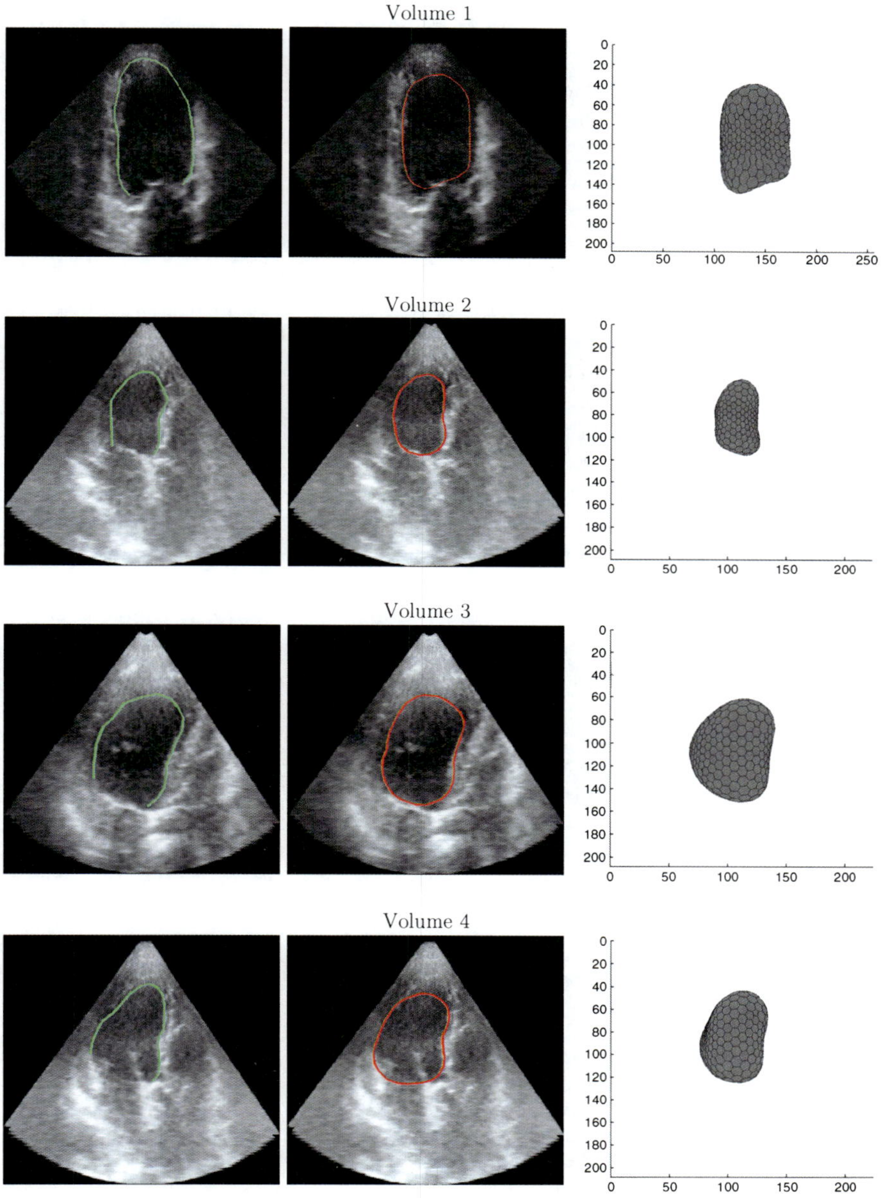

Fig. 6 Slice from the echocardiographic showing (on the *left*) the GT and (on the *center*) the obtained segmentation. Final configuration of the surface (on the *right*)

Table 2 Results of the evaluation methods for each volume ($\bar{d}_{AV}$, $\bar{d}_{HDF}$ and $\bar{d}_{MAD}$ are expressed in millimeters)

	Volume 1	Volume 2	Volume 3	Volume 4
$\bar{d}_{HMD}$	0.15 ± 0.03	0.20 ± 0.10	0.16 ± 0.02	0.24 ± 0.04
$\bar{d}_{AV}$	3.1 ± 0.6	2.5 ± 1.1	2.7 ± 0.5	3.8 ± 0.6
$\bar{d}_{HDF}$	9.2 ± 2.1	7.2 ± 2.6	6.9 ± 1.7	11.3 ± 1.9
$\bar{d}_{MAD}$	4.5 ± 1.1	3.9 ± 2.3	3.6 ± 0.9	6.7 ± 2.1

8 Conclusions

This paper addresses the automatic LV segmentation problem in 3D echocardiographic data. Due to the nature of the volumes, many of the detected features are outliers and do not belong to the LV boundary. The proposed system uses a robust estimation technique based on PDAF that prevents the segmentation to be misguided by the outliers and leads to acceptable surface estimates.

The results show that the proposed system performs a good segmentation of the LV. It achieves an average error of 3 mm between the obtained segmentation and the GT, which is within the state-of-the-art results [43]. Therefore, this system has the potential to be used to accurately compute cardiac measurements such the systemic and diastolic volumes.

Further tests shows that the developed system is still over-dependent on the initialization procedure, which does not help improving the repeatability of LV segmentations. This weakness could be alleviated using an automatic initialization scheme, such as in [42].

Acknowledgements This work was supported by project [PTDC/EEA-CRO/103462/2008] (project HEARTRACK) and FCT [PEst-OE/EEI/LA0009/2011].

References

1. Barber, C.B., Dobkin, D.P., Huhdanpaa, H.: The quickhull algorithm for convex hulls. ACM Trans. Math. Software **22**, 469–483 (1996)
2. Bar-Shalom, Y., Fortmann, T.: Tracking and Data Associaton. Academic, New York (1988)
3. Blake, A., Isard, M.: Active Contour. Springer, Berlin (1998)
4. Bernard, O., Touil, B., Gelas, A., Prost, R., Friboulet, D.: A rbf-based multiphase level set method for segmentation in echocardiography using the statistics of the radiofrequency signal, IEEE International Conference on Image Processing (ICIP) 3, III-157-III-160 (2007)
5. Bernard, O., Friboulet, D., Thevenaz, P., Unser, M.: Variational b-spline level-set: A linear filtering approach for fast deformable model evolution. IEEE Trans. Imag. Proc. **18**(6), 1179–1991 (2009)
6. Bosch, J.G., Mitchell, S.C., Lelieveldt, B.P.F., Nijland, F., Kamp, O., Sonka, M., Reiber, J.H.C.: Automatic segmentation of echocardiographic sequences by active appearance motion models. IEEE Trans. Med. Imag. **21**(11), 1374–1383 (2002)
7. Bosch, J.G., Mitchell, S.C., Lelieveldt, B.P.F., Nijland, F., Kamp, O., Sonka, M., Reiber, J.H.C.: Automatic segmentation of echocardiographic sequences by active appearance motion models. IEEE Trans. Med. Imag. **21**(11), 1374–1383 (2002)

8. Carneiro, G., Nascimento, J.C.: Multiple dynamic models for tracking the left ventricle of the heart from ultrasound data using particle filters and deep learning architectures, IEEE Conference on Computer Vision and Pattern Recognition (CVPR), 2815–2822 (2010)
9. Carneiro, G., Georgescu, B., Good, S., Comaniciu, D.: Detection and measurement of fetal anatomies from ultrasound images using a constrained probabilistic boosting tree. IEEE Trans. Med. Imag. **27**(9), 1342–1355 (2008)
10. Chan, K., Veinot, J.P.: Anatomic Basis of Echocardiographic Diagnosis, 1st edn. Oxford Univesity Press, Oxford (2011)
11. Chen, T., Babb, J., Kellman, P., Axel, L., Kim, D.: Semiautomated segmentation of myocardial contours for fast strain analysis in cine displacement-encoded MRI. IEEE Trans. Med. Imag. **27**(8), 1084–1094 (2008)
12. Cootes, T., Taylor, C., Cooper, D., Graham, J.: Active shape models – their training and application. Comput. Vis. Image Understand. **61**(1), 38–59 (1995)
13. Cootes, T., Beeston, C., Edwards, G., Taylor, C.: A unified framework for atlas matching using active appearance models, Information Processing in Medical Imaging, Lecture Notes in Computer Science 1613/1999, 322–333 (1999)
14. Cremers, D., Osher, S., Soatto, S.: Kernel density estimation and intrinsic alignment for shape priors in level set segmentation. Int. J. Comput. Vision **69**(3), 335–351 (2006)
15. Debreuve, E., Barlaud, M., Aubert, G., Laurette, I., Darcourt, J.: Space-time segmentation using level set active contours applied to myocardial gated SPECT. IEEE Trans. Med. Imag. **20**(7), 643–659 (2001)
16. Delingette, H.: General Object Reconstruction Based on Simplex Meshes. In: International Journal of Computer Vision, **32**(2), 111–146 (1999)
17. Duan, Q., Angelini, E.D., Laine, A.: Real time segmentation by active geometric functions. Comput. Meth. Prog. Biomed. **98**(3), 223–230 (2010)
18. Freund, Y., Schapire, R.: A decision-theoretic generalization of on-line learning and an application to boosting. J. Comput. Syst. Sci. **55**(1), 119–139 (1997)
19. Georgescu, B., Zhou, X.S., Comaniciu, D., Gupta, A.: Database-guided segmentation of anatomical structures with complex appearance. In: Conference on Computer Vision and Pattern Rec (CVPR) (2005)
20. Hammoude, A.: Endocardial border identification in two-dimensional echocardiographic images: review of methods. Comput. Med. Imag. Graph. **22**(3), 181–193 (1998)
21. Hang, X., Greenberg, N.L., Thomas, J.D.: Left ventricle quantification in 3D echocardiography using a geometric deformable model. In: Computers in Cardiology, **31**, 649–652 (2004)
22. Yu, H., Pattichis, M.S., Goens, M.B.: Robust Segmentation and Volumetric Registration in a Multi-view 3D Freehand Ultrasound Reconstruction System, Fortieth Asilomar Conference on Signals, Systems and Computers (ACSSC), 1978–1982 (2006)
23. Huttenlocher, D.P., Ullman, S.: Recognizing solid objects by alignment with an image. Int. J. Comput. Vision **5**(2) (1990)
24. Juang, R., McVeigh, E., Hoffmann, B., Yuh, D., Burlina, P.: Automatic segmentation of the left-ventricular cavity and atrium in 3d ultrasound using graph cuts and the radial symmetry transform. IEEE International Symposium on Biomedical Imaging: From Nano to Macro, 606–609 (2011)
25. Kass, M., Witkin, A., Terzopoulos, D.: Snakes: Active contour models. Int. J. Comput. Vision **4**(1), 321–331 (1987)
26. Kutulakos, K.N., Seitz, S.M.: A theory of shape by space carving. In: Seventh IEEE International Conference on Computer Vision, **1**, 307–314 (1999)
27. Lin, N., Yu, W., Duncan, J.: Combinative multi-scale level set framework for echocardiographic image segmentation. Med. Imag. Anal. **7**(4), 529–537 (2003)
28. Lynch, M., Ghita, O., Whelan, P.F.: Segmentation of the left ventricle of the heart in 3-D+t MRI data using an optimized nonrigid temporal model. IEEE Trans. Med. Imag. **27**(2), 195–203 (2008)
29. Malladi, R., Sethian, J., Vemuri, B.: Shape modeling with front propagation: A level set approach. IEEE Trans. Pattern Anal. Mach. Intell. **17**, 158–175 (1995)

30. Mitchell, S., Lelieveldt, B., van der Geest, R., Bosch, H., Reiber, J., Sonka, M.: Multistage hybrid active appearance model matching: Segmentation of left and right ventricles in cardiac MR images. IEEE Trans. Med. Imag. **20**(5), 415–423 (2001)
31. Nascimento, J.C., Marques, J.S.: Robust shape tracking in the presence of cluttered background. In: IEEE Transactions on Multimedia, **6**(6), 852–861 (2004)
32. Nascimento, J.C., Marques, J.S.: Robust shape tracking with multiple models in ultrasound images. IEEE Trans. Imag. Proc. **17**(3), 392–406 (2008)
33. Nillesen, M.M., Lopata, R.G.P., Gerrits, I.H., Kapusta, L., Huisman, H.J., Thijssen, J.M., de Korte, C.L.: 3D Segmentation of the Heart Muscle in Real-Time 3D Echocardiographic Sequences Using Image Statistics. In: IEEE Ultrasonics Symposium, 1987–1990 (2006)
34. Noble, J.A., Boukerroui, D.: Ultrasound image segmentation: A survey. IEEE Trans. Med. Imag. **25**(8), 987–1010 (2006)
35. Orderud, F., Hansgård, J., Rabben, S.I.: Real-time tracking of the left ventricle in 3d echocardiography using a state estimation approach. In: Proceedings of the 10th International Conference on Medical Image Computing and Computer-Assisted Intervention - Volume Part I, MICCAI'07, pp. 858–865. Springer, Berlin (2007)
36. Paragios, N.: A level set approach for shape-driven segmentation and tracking of the left ventricle. IEEE Trans. Med. Imag. **22**(6), 773–776 (2003)
37. Paragios, N., Deriche, R.: Geodesic active regions and level set methods for supervised texture segmentation. Int. J. Comput. Vision **46**(3), 223–247 (2002)
38. Santiago, C., Marques, J.S., Nascimento, J.C.: Robust Deformable Model for Segmenting the Left Ventricle in 3D Volumes of Ultrasound Data. In: International Conference on Pattern Recognition Applications and Methods, (2012) (to be published by Elsevier in ICPRAM'12 February)
39. Sarti, A., Corsi, C., Mazzini, E., Lamberti, C.: Maximum likelihood segmentation of ultrasound images with rayleigh distribution. IEEE Trans. Ultrason. Ferroelect. Frequen. Contr. **52**(6), 947–960 (2005)
40. Scowen, B., Smith, S., Vannan, M.: Quantitative 3d modelling of the left ventrical from ultrasound images. Euromicro Conf. **2**, 432–439 (2000)
41. Sonka, M., Zhang, X., Siebes, M., Bissing, M., Dejong, S., Collins, S., Mckay, C.: Segmentation of intravascular ultrasound images: A knowledge-based approach. IEEE Trans. Med. Imag. **14**, 719–732 (1995)
42. Yang L., Georgescu, B., Zheng, Y., Meer, P., Comaniciu, D.: 3D ultrasound tracking of the left ventricle using one-step forward prediction and data fusion of collaborative trackers. In: IEEE Conference on Computer Vision and Pattern Recognition, 1–8 (2008)
43. Yang, L., Georgescu, B., Zheng, Y., Wang, Y., Meer, P., Comaniciu, D.: Prediction based collaborative trackers (pct): A robust and accurate approach toward 3d medical object tracking. IEEE Trans. Med. Imag. **30**, 1921–1932 (2011)
44. Zagrodsky, V., Walimbe, V., Castro-Pareja, C., Qin, J.X., Song, J.-M., Shekhar, R.: Registration-assisted segmentation of real-time 3-D echocardiographic data using deformable models. IEEE Trans. Med. Imag. **24**(9), 1089–1099 (2005)
45. Zhang, L., Geiser, E.: An effective algorithm for extracting serial endocardial borders from 2-d echocardiograms. IEEE Trans. Biomed. Eng. **BME-31**, 441–447 (1984)
46. Zhou, X.S., Comaniciu, D., Gupta, A.: An information fusion framework for robust shape tracking. IEEE Trans. Pattern Anal. Mach. Intell. **27**(1), 115–129 (2005)

Facial Expression Recognition Using Diffeomorphic Image Registration Framework

Bartlomiej W. Papiez, Bogdan J. Matuszewski, Lik-Kwan Shark, and Wei Quan

Abstract This paper presents a new method for facial expression modelling and recognition based on diffeomorphic image registration parameterised via stationary velocity fields in the log-Euclidean framework. The validation and comparison are done using different statistical shape models (SSM) built using the Point Distribution Model (PDM), velocity fields and deformation fields. The obtained results show that the facial expression representation based on stationary velocity fields can be successfully utilised in facial expression recognition, and this parameterisation produces a higher recognition rate than the facial expression representation based on deformation fields.

Keywords Facial expression representation • Facial expression recognition • Vectorial log-Euclidean statistics • Statistical shape modelling • Diffeomorphic image registration

1 Introduction

Face is an important medium used by humans not only to communicate, but also reflecting a person's emotional and awareness states, cognitive activity, personality or well-being. Over the last 10 years automatic facial expression representation and recognition have become an area of significant research interest for the computer vision community, with applications in human–computer interaction (HCI) systems, medical/psychological sciences and visual communications to name a few.

B.W. Papiez (✉) • B.J. Matuszewski • L.-K. Shark • W. Quan
Applied Digital Signal and Image Processing Research Centre, University of Central Lancashire, PR1 2HE Preston, UK
e-mail: bartlomiej.papiez@eng.ox.ac.uk; bmatuszewski1@uclan.ac.uk; lshark@uclan.ac.uk; wquan@uclan.ac.uk

P. Latorre Carmona et al. (eds.), *Mathematical Methodologies in Pattern Recognition and Machine Learning*, Springer Proceedings in Mathematics & Statistics 30,
DOI 10.1007/978-1-4614-5076-4_12,

Although significant efforts have been undertaken to improve the facial features extraction process and the recognition performance, automatic facial expression recognition is still a challenging task due to an inherent subjective nature of the facial expressions and their variation over different gender, age and ethnicity groups. Detailed overviews of existing methodologies, recent advances and challenges can be found in [7, 12, 14, 23].

The facial expression representation can be seen as a process of extracting features, which could be generic such as local binary patterns [21] or Gabor coefficients [3], or more specific such as landmarks of characteristic points located in areas of major facial changes due to articulation [11], or a topographic context (TC) that treats the intensity levels of an image as a 3-D terrain surface [26]. Recently, in [18, 19] authors postulated that the space shape vectors (SSV) of the statistical shape model (SSM) can constitute a significant feature space for the recognition of facial expressions. The SSM can be constructed in many different ways, and it was developed based on the point distribution model originally proposed by [6]. In [17], the SSM is built based on the control points of the B-Spline surface of the training data set, and in [20] an improved version with multi-resolution correspondence search and multi-level model deformation was proposed. In this paper, the SSM is generated using the stationary velocity fields obtained from diffeomorphic face registration [16]. The idea of using the motion fields as feature in computer vision and pattern recognition was used previously for face recognition where the optical flow was computed to robustly recognise face with different expressions based on a single sample per class in the training set [10].

In medical image analysis, the parameterisation of the diffeomorphic transformation based on the principal logarithm of non-linear geometrical deformations was introduced in [1]. Using this framework, the log-Euclidean vectorial statistics can be performed on the diffeomorphic vector fields via their logarithm, which always preserve the invertibility constraint contrary to the Euclidean statistics on the deformation fields. Recently, the stationary velocity field parametrisation has been utilised for deformable image registration in different ways e.g. for exponential updates of deformation field [25], or producing the principal logarithm directly as an output of image registration e.g. inverse consistent image registration [2, 24] or symmetric inverse consistent image registration [9]. These algorithms preserve the spatial topology of objects by maintaining diffeomorphism. As the facial shapes (mouth, eyes, eye brows) have constant intra- and inter-subject topology, it is interesting to check the adequacy of the facial expressions represented using stationary velocity fields as a result of performing diffeomorphic image registration and compare with the deformation field-based facial expression representation in terms of separability in feature space and recognition performance.

The remainder of the paper is organised as follows: Sect. 2 introduces the concept of the SSM with detailed description of the group-wise registration algorithm (Sect. 2.1). Then, the velocity field-based representation of facial expression is described in Sect. 2.2, and the Point Distribution Model is presented in Sect. 2.3. The experimental results of qualitative and quantitative evaluation are shown in Sect. 3 with concluding remarks in Sect. 4.

2 Log-Euclidean Statistical Shape Model

The statistical shape model was developed based on the point distribution model originally proposed in [6]. The model represents the facial expression variations based on the statistics calculated for corresponding features during the learning process of the training data set. In order to build an SSM, the correspondence of facial features between different faces in the training data set must be established. This is done here first by generating a *mean* face model for the neutral facial expression data set to find the mappings from any face to the so-called *common face space*. Then, by transferring subject-specific facial expressions data set into the common face space, the intra-subject facial expression correspondence is estimated. Finally, the principal component analysis (PCA) is applied to the training data set aligned in the common face space to provide a low-dimensional feature space for facial expression representation.

2.1 *Log-Domain Group-wise Image Registration*

Generation of the *mean* face model is an essential step during the training process because it allows a subject-independent common face space to be established for further analysis.

For a given set of n-dimensional images representing neutral facial expressions denoted by

$$\mathbf{I}^{\text{ne}} = \{I_k^{\text{ne}} : \Omega \subset \mathbb{R}^n \to \mathbb{R}, k = 1, \ldots, K\} \tag{1}$$

where K is the number of subjects included in training data, and the objective is to estimate a set of displacement fields $\hat{\mathbf{u}}^{\text{ne}}$ to map the images taken from $\mathbf{I}^{\text{ne}}$ to the *mean* face model I^{mean}.

In general, this problem can be formulated as a minimisation problem:

$$\hat{\mathbf{u}}^{\text{ne}} = \arg\min_{\mathbf{u}^{\text{ne}}} \epsilon(\mathbf{u}^{\text{ne}}; \mathbf{I}^{\text{ne}}) \tag{2}$$

where $\epsilon(\mathbf{u}^{\text{ne}})$ is defined as

$$\begin{aligned} \epsilon(\mathbf{u}^{\text{ne}}) = & \sum_k \sum_l \int_\Omega \text{Sim}(I_k^{\text{ne}}(x + u_k(x)), I_l^{\text{ne}}(x + u_l(x)))\text{d}x \\ & + \alpha \sum_k \int_\Omega \text{Reg}(u_k(x))\text{d}x \end{aligned} \tag{3}$$

where $x = [x^1, \ldots, x^n] \in \Omega$ denotes given voxel position, Sim denotes a similarity measure between each pair of the images, I_k^{ne} and I_l^{ne} $(l \neq k)$ from $\mathbf{I}^{\text{ne}}$, Reg denotes a regularisation term, and α is a weight of the regularisation term. In this work, the deformation fields are parameterised by recently proposed stationary velocity fields $v(x)$ via exponential mapping [1]:

$$\varphi(x) = x + u(x) = x + \exp(v(x)). \tag{4}$$

To minimise (2), Demon force [25] was used in the symmetric manner [15] in the following way:

$$\mathrm{d}u_{kl}^{i} = \frac{(I_k^{\varphi_k^i} - I_l^{\varphi_l^i})(\nabla I_k^{\varphi_k^i} + \nabla I_l^{\varphi_l^i})}{\|\nabla I_k^{\varphi_k^i} + \nabla I_l^{\varphi_l^i}\|^2 + (I_k^{\varphi_k^i} - I_l^{\varphi_l^i})^2} \tag{5}$$

where $I_k^{\varphi_k^i} = I_k^{\mathrm{ne}}(\varphi_k^i(x))$, $I_l^{\varphi_l^i} = I_l^{\mathrm{ne}}(\varphi_l^i(x))$ are warped images and $\nabla I_k^{\varphi_k^i}$, $\nabla I_l^{\varphi_l^i}$ are gradients of those images, and i is an iteration index. The average update of the velocity field is calculated using the log-Euclidean mean for vector fields $\mathrm{d}u_{kl}^{i}$ given by [1]:

$$\mathrm{d}v_k^i = \frac{1}{K}\sum_l \log(\mathrm{d}u_{kl}^i) \tag{6}$$

and the deformation field $u_k^{i+1}(x)$ is calculated via exponential mapping for the updated velocity field:

$$v_k^{i+1}(x) = v_k^i(x) + \mathrm{d}v_k^i(x) \tag{7}$$

Although according to (6) the log-Euclidean mean requires calculation of the logarithm, which is reported to be a time-consuming process [1, 5], the consistent log-domain diffemorphic Demon approach [24] is used which produces the principal logarithm of transformation as an output of image registration and therefore the logarithm is not calculated directly. Finally, the *mean* face model is generated by averaging the intensity of all images after registration:

$$I^{\mathrm{mean}} = \frac{1}{K}\sum_k^K I_k^{\mathrm{ne}}(\varphi_k(x)) \tag{8}$$

The procedure for estimation of the set of deformation fields for generation of the common face space is summarised below [16]:

```
repeat
   for k=1:K
      for l=1:K and l!=k
          Calculate update (Equation 5)
      end
      Calculate average of updates (Equation 6)
      Update velocity field (Equation 7)
      Smooth velocity field using Gaussian filter
   end
   i = i+1;
until (velocity fields do not change) or
      (i>max_Iteration)
```

The examples of the *mean* face model estimated by applying the proposed scheme to neutral expressions are illustrated in Fig. 1, where different input data sets are used for validation. Moreover, the quantitative performance of the proposed implicit group-wise registration includes the Intensity Variance (IV) criterion [22]. The intensity variance measures the similarity of the group of images (population) based on the pixel intensity differences. The IV is computed here as follows:

$$IV(x) = \frac{1}{K-1}\sum_{k}^{K}(I_k^{\mathrm{ne}}(\varphi_k(x)) - I^{\mathrm{mean}}(x))^2 \tag{9}$$

where I^{mean} is given by (8). The perfect group-wise registration results for the images of the same modality should be characterised with the minimum pixel intensity differences between registered images.

The presented algorithm of generating the *mean* face model is similar to the work presented in [8]. The main difference is in how the deformation fields are parameterised with the stationary velocity field used in the proposed method instead of the Fourier series in [8], and secondly in the method of solving (2) with the Demon approach used instead of the linear elastic model. Using the log-domain parameterisation for deformation fields is reported to produce smoother deformation fields and it allows vectorial statistics to be calculated directly on the velocity fields.

2.2 *Velocity Field-Based Facial Expression Model*

The next step is to warp all other training faces representing different facial expressions to the mean face (the reference face) via transformation $\varphi_k(x)$ estimated for neutral expressions. For a given set of facial expression images from subject K:

$$\mathbf{I}_{\mathbf{k}}^{\mathrm{ex}} = \{I_{km}^{\mathrm{ex}} : \Omega \subset \mathbb{R}^n \rightarrow \mathbb{R}, m = 1, \ldots, M\} \tag{10}$$

where M denotes the number of images. The transformation $\varphi_k(x)$ is applied to get a set of facial expression images in the common face space (space of the reference image):

$$\mathbf{I}_{\mathbf{k}}^{\mathrm{cex}} = \{I_{km}^{\mathrm{ex}}(\varphi_k(x))\} \tag{11}$$

By applying the log-Domain image registration approach based on the consistent Demon algorithm [24], each image in set $\mathbf{I}_{\mathbf{k}}^{\mathrm{cex}}$ is registered to image of neutral expression in common face space $I_k^{\mathrm{ne}}(\varphi_k(x))$, the set of the velocity fields $\mathbf{v}_{\mathbf{k}}^{\mathrm{ex}}$ is estimated, and the set of the corresponding deformation fields $\mathbf{u}_{\mathbf{k}}^{\mathrm{ex}}$ via exponential mapping is calculated as well. Utilising this particular method for image registration has two important advantages. Firstly, the consistency criterion is maintained during the registration process that helps to keep the smooth transformation especially for cases like matching between open-mouth and close-mouth shapes. Secondly, the results of registration are the velocity fields so there is no necessity of calculating the principal logarithm of transformations.

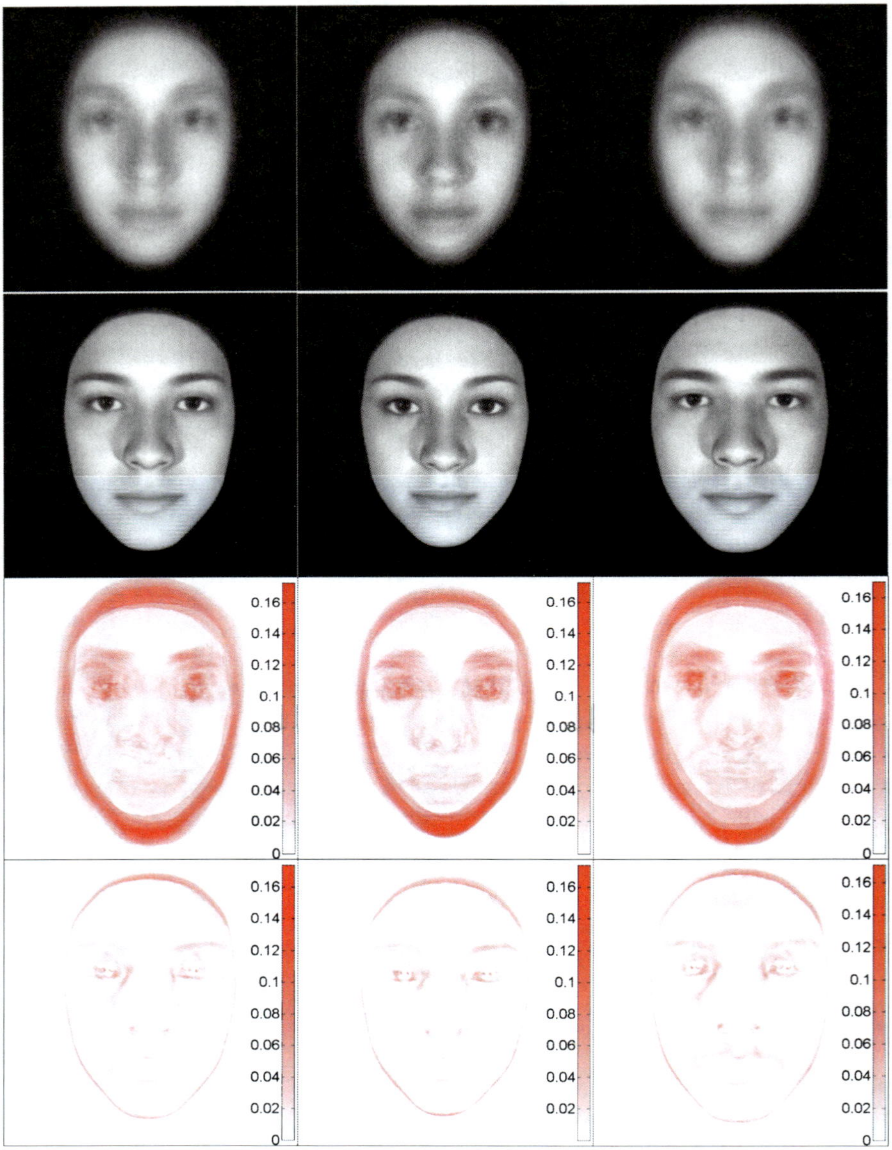

Fig. 1 Grey-level average of mean face before registration, and after registration: (from *left* to *right*) mixed data set, female data set, and male data set. (*Top row*) shows grey-level average before registration, (*upper middle row*) shows grey-level average of images after registration, (*lower middle row*) shows the *IV* before registration and (*bottom row*) shows the *IV* after registration

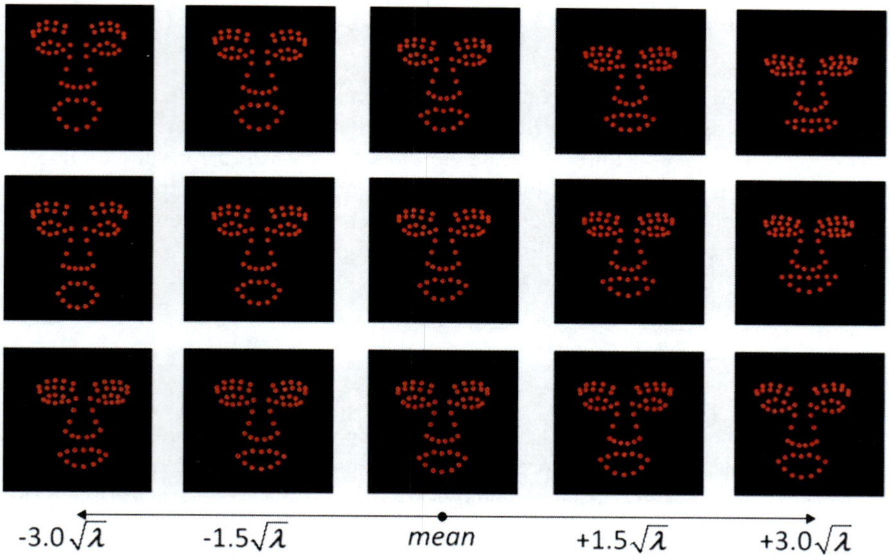

Fig. 2 Variations of the first (*top row*), the second (*middle row*), and the third (*bottom row*) major modes of the point distribution model for automatically selected landmarks

2.3 *Point Distribution Model*

The point distribution model originally proposed by [6] is one of the most often used techniques for representing shapes. This model describes a shape as a set of positions (landmarks) in the reference image. The variations between different shapes require establishment of the correspondence between points detected in the reference image and images representing different deformation in the training set. Although this can be relatively reliably achieved during the model training phase by careful time-consuming, often manual selection of corresponding points, such task is prone to occurrence of gross errors during the model evaluation where often near real-time performance is required. The examples of the manually selected landmarks for expressions included in data sets can be found in [16, 27]. The automatically selected landmarks used later on in the experimental section are obtained with the help of face image registration described in the previous section. In that case, the manually selected landmarks in the model face are automatically mapped into registered faces.

Using the standard principal component analysis (PCA), each face representation in the training data set can be approximately represented in a low-dimensional shape vector space instead of the original high-dimensional data vector space [4]. Figure 2 shows the effect of varying the first three largest principal components of the PDM for automatically selected landmarks, where λ is eigenvalue of the covariance matrix calculated from the training data set.

3 Experimental Results

The data set used for validation consists of 48 subjects that are selected from the BU-3DFE database [27], with a wide variety of ethnicity, age and gender. The data used during the training procedure are excluded from the data used for validation. The implicit group-wise registration based on the Demon force minimises the Sum of Squared Difference (SSD) between images and hence due to different skin patterns an additional image intensity adjustment was performed.

3.1 *Separability Analysis*

To assess whether the Shape Space Vectors based on the velocity fields can be used as a feature space for facial expression analysis and recognition, the separability of the SSV-based features has been analysed. The first three elements of the SSM are used to reveal clustering characteristics and separability powers. The SSM for training was built using 24 subjects, each containing 25 faces, the SSV is based on the automatically selected points (with 60 landmarks per face), the velocity fields, and the deformation fields (with 512×512 pixels per image). The test data set was extracted from another 24 subjects. The training data set and the testing data set are mutually exclusive. Examples for some pairs of the expressions given in Fig. 3 exhibit good separability even in the low-dimensional space, especially for expressions such as “happiness vs. sadness” or “disgust vs. surprise”. The expressions like “anger vs. fear” appear to overlap more with each other, but the clusters can be identified.

In order to quantitatively assess the separability of the presented facial expression features, the appropriate criteria are needed to be calculated. A computable criterion for measurement of within-class and between-class distances was applied in similar way as it was done by [19, 26]. The within-class scatter matrix S_W is defined as follows:

$$S_W = \sum_{i=1}^{c} \frac{1}{n} \sum_{k=1}^{n_i} (x_k^i - m_i)(x_k^i - m_i)^{\mathrm{T}} \tag{12}$$

and the between-class scatter matrix S_B is defined as:

$$S_B = \sum_{i=1}^{c} \frac{n_i}{n} (m_i - m)(m_i - m)^{\mathrm{T}} \tag{13}$$

where: x_k^i is a d-dimensional feature, n_i is the number of samples in the ith class, n is the number of samples in all classes, c is the number of classes, m_i is the mean of samples in the ith class defined as:

$$m_i = \frac{1}{n_i} \sum_{k=1}^{n_i} x_k^i \tag{14}$$

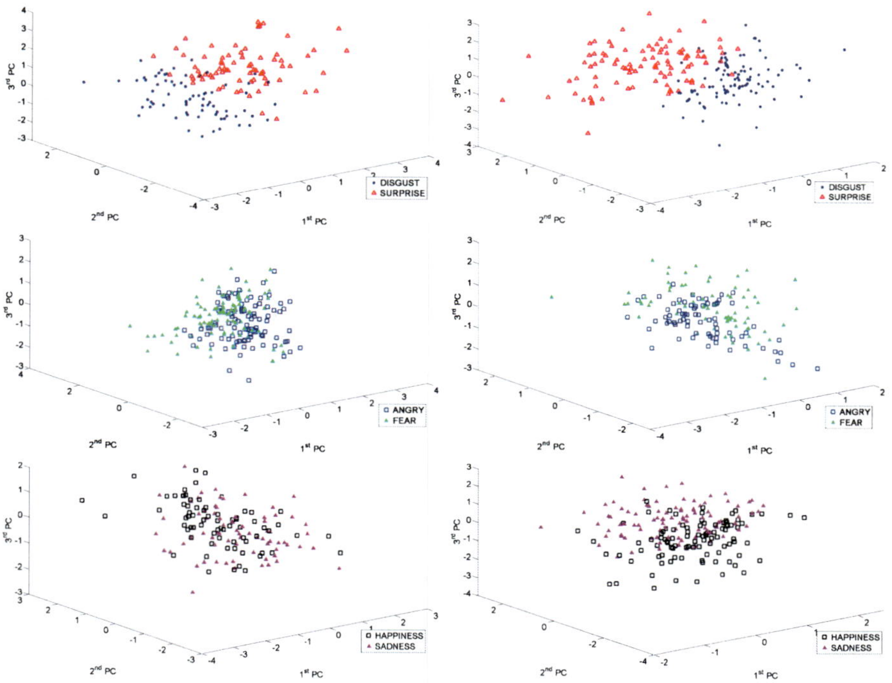

Fig. 3 Separability analysis for automatically selected landmarks (*left*) and for full velocity field (*right*) using first three principal components

m is the mean of all the samples:

$$m = \sum_{i=1}^{c} \frac{n_i}{n} m_i \tag{15}$$

The separability criterion $J_2(x)$ is defined as a natural logarithm of the ratio of within-class scatter matrix's determinant and between-class scatter matrix's determinant:

$$J_2(x) = \ln \frac{\det(S_B + S_W)}{\det(S_W)} \tag{16}$$

This separability criterion is efficient for comparison of different feature selection, lying in the completely different spaces, and it is intrinsically normalised and reflects the quantity of separability for features between different classes [19,26]. The larger value of $J_2(x)$ means the better separability.

The separability criterion was evaluated on the different facial expression representations, namely the manually selected landmarks, the automatically detected landmarks, the full velocity fields, and the full deformation fields, and the results are shown in Fig. 4. The results can be summarised that for the same ratio of retained

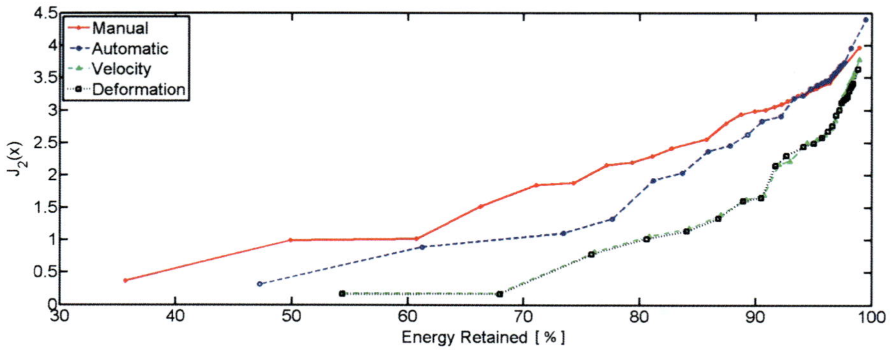

Fig. 4 Separability of expression for different features in term of separability criterion $J_2(x)$

energy in the training data, the value of $J_2(x)$ for the manually selected landmarks is the highest. The automatically selected landmarks in range above 80% is not significantly different from the manually selected landmarks. The velocity field and the deformation field-based facial expression representations are the worst.

To quantify the between-expression separability, the two-class separability criterion was evaluated [26]. The within-class scatter matrix $S_W^{\mathrm{ex}_i,\mathrm{ex}_j}$ for two-class case $(c = 2)$ is defined as follows:

$$S_W^{\mathrm{ex}_i,\mathrm{ex}_j} = \frac{1}{n}\left(\sum_{k=1}^{n_{\mathrm{ex}_i}}(x_k^{\mathrm{ex}_i} - m_{\mathrm{ex}_i})(x_k^{\mathrm{ex}_i} - m_{\mathrm{ex}_i})^{\mathrm{T}} + \sum_{l=1}^{n_{\mathrm{ex}_j}}(x_l^{\mathrm{ex}_j} - m_{\mathrm{ex}_j})(x_l^{\mathrm{ex}_j} - m_{\mathrm{ex}_j})^{\mathrm{T}}\right) \quad (17)$$

and the between-class scatter matrix $S_B^{\mathrm{ex}_i,\mathrm{ex}_j}$ is defined:

$$S_B^{\mathrm{ex}_i,\mathrm{ex}_j} = \frac{n_{\mathrm{ex}_i}n_{\mathrm{ex}_j}}{n^2}(m_{\mathrm{ex}_i} - m_{\mathrm{ex}_j})(m_{\mathrm{ex}_i} - m_{\mathrm{ex}_j})^{\mathrm{T}} \quad (18)$$

where ex_i and ex_j are analysed expressions, n_{ex_i}, n_{ex_j} are the numbers of samples in the ith and jth class, $n = n_{\mathrm{ex}_i}+n_{\mathrm{ex}_j}$. For each pair of selected expressions $J_2^{\mathrm{ex}_i,\mathrm{ex}_j}(x)$ of different facial expression representation was calculated.

Tables 1–4 shows the separability of all pairs of expression for different facial expression representations. These results support the visual inspection of the qualitative analysis presented in Fig. 3. The separability of the pair of expressions such as "happiness and sadness", or "disgust and surprise" gets higher values of separability criterion $J_2^{\mathrm{ex}_i,\mathrm{ex}_j}(x)$ (with the minimum value of 2.57), while the separability of the pair of "anger and fear" is lower (with the maximum value of 2.36).

Table 1 Confusion matrix of $J_2^{\mathrm{ex}_i,\mathrm{ex}_j}(x)$ for the manually selected landmarks

	Anger	Disgust	Fear	Happiness	Sadness	Surprise
Anger	–	2.15	2.36	3.71	1.70	4.10
Disgust	–	–	2.43	3.38	2.82	3.27
Fear	–	–	–	2.09	2.05	2.78
Happiness	–	–	–	–	3.95	4.44
Sadness	–	–	–	–	–	3.90
Surprise	–	–	–	–	–	–

Table 2 Confusion matrix of $J_2^{\mathrm{ex}_i,\mathrm{ex}_j}(x)$ for the automatically detected landmarks

	Anger	Disgust	Fear	Happiness	Sadness	Surprise
Anger	–	2.11	2.15	3.29	1.57	3.34
Disgust	–	–	2.23	3.31	2.45	3.06
Fear	–	–	–	2.06	1.92	2.51
Happiness	–	–	–	–	3.52	4.02
Sadness	–	–	–	–	–	3.20
Surprise	–	–	–	–	–	–

Table 3 Confusion matrix of $J_2^{\mathrm{ex}_i,\mathrm{ex}_j}(x)$ for the full deformation fields

	Anger	Disgust	Fear	Happiness	Sadness	Surprise
Anger	–	1.81	2.21	2.56	1.51	4.12
Disgust	–	–	2.14	2.19	2.29	3.19
Fear	–	–	–	1.61	1.68	2.69
Happiness	–	-	–	–	2.57	3.40
Sadness	–	–	–	–	–	3.41
Surprise	–	–	–	–	–	–

Table 4 Confusion matrix of $J_2^{\mathrm{ex}_i,\mathrm{ex}_j}(x)$ for the full velocity fields

	Anger	Disgust	Fear	Happiness	Sadness	Surprise
Anger	–	1.91	2.21	2.61	1.52	3.94
Disgust	–	–	2.14	2.20	2.32	3.14
Fear	–	–	–	1.62	1.71	2.68
Happiness	–	–	–	–	2.61	3.42
Sadness	–	–	–	–	–	3.46
Surprise	–	–	–	–	–	–

3.2 Experiments on Facial Expression Recognition

The separability analysis conducted in the previous section indicates that the SSV feature space based on the velocity can be used for classification of facial expressions. Data used for classification-based validation again consists of 48 subjects, and contains neutral expression and six basic facial expressions of anger, disgust, fear, happiness, sadness and surprise with four different expression intensity ranges. These data were divided into six subsets containing 8 subjects with 25

Table 5 Recognition rate for different classifier methods

Feature/ classifier	LDA (%± SD)	QDA (%± sSD)	NNC (%± SD)	NBC (%± SD)
Manually	78.1±4.2	74.0±4.8	61.5±1.1	74.3±2.4
Automatic	73.4±6.0	**69.1±6.0**	**61.9±4.9**	**70.8±4.0**
Deformation	75.0±5.2	58.9±2.9	56.2±5.0	69.1±5.0
Velocity	**76.6±4.3**	59.3±4.1	57.7±5.1	69.1±5.2

Table 6 Confusion matrix of the LDA for the manually selected landmarks

Input/ Output	Anger (%)	Disgust (%)	Fear (%)	Happiness (%)	Sadness (%)	Surprise (%)
Anger	**74.5**	4.7	3.1	3.1	14.6	0.0
Disgust	8.3	**81.8**	4.7	0.5	3.6	1.0
Fear	7.8	1.6	**59.9**	11.5	16.1	3.1
Happiness	4.2	2.1	8.3	**85.4**	0.0	0.0
Sadness	16.7	1.6	4.2	0.0	**77.6**	0.0
Surprise	1.0	2.1	4.2	0.5	2.6	**89.6**

faces per subject representing different expressions. During evaluation one subset is chosen as the testing set, and the remaining data are used for the training, and the evaluation procedure is repeated six times, every time with the different testing set. Four types of facial expression representations have been used for validation: the manually selected landmarks from the database [27], the automatically detected facial landmarks using the log-domain Demon registration, the full velocity fields, and the full deformation fields.

Four commonly used classification methods were used for evaluation, namely linear discriminant analysis (LDA), quadratic classifier (QDC), nearest neighbour classifier (NCC), and naive Bayes classifier (NBC). The detailed description of these methods can be found in most of the textbooks on pattern recognition e.g. [4].

The average recognition rates and standard deviations of all six experiments for different facial expression data are presented in Table 5. It can be seen that the LDA classifier achieves the highest recognition rate for every facial expression representation. As shown in Table 5 all facial expression representations achieve a similar recognition rate for the same classifier with the highest rate for the manually selected landmarks. The manually selected landmarks are included only for a reference for other automatic methods. The recognition rates obtained by the automatic methods are lower (maximum 15.1 % less using the deformation field-based representation and QDA classifier) than that obtained by manual landmark selection. The confusion matrices for LDA for different data are given in Tables 6–9. From the classification performance, it can be concluded that the surprise, disgust, happiness and sadness expressions can be classified in most cases with above 75 % accuracy, anger with about 70 % accuracy, whereas fear is only classified correctly in 61.5 % using the velocity field-based representation. The best recognition rates (about 90 %) are found for surprise, similar to the work reported in [19] for data sets taken from the same database.

Table 7 Confusion matrix of the LDA for the automatically detected landmarks

Input/ Output	Anger (%)	Disgust (%)	Fear (%)	Happiness (%)	Sadness (%)	Surprise (%)
Anger	**68.8**	5.2	5.2	2.6	18.2	0.0
Disgust	12.5	**76.6**	5.7	0.5	3.6	1.0
Fear	7.8	2.6	**55.2**	14.1	19.3	1.0
Happiness	4.1	1.6	11.5	**82.3**	0.0	0.5
Sadness	19.8	3.1	4.7	0.0	**72.4**	0.0
Surprise	1.0	3.1	7.8	0.5	2.6	**87.0**

Table 8 Confusion matrix of the LDA for the full deformation fields

Input/ Output	Anger (%)	Disgust (%)	Fear (%)	Happiness (%)	Sadness (%)	Surprise (%)
Anger	**74.5**	9.9	1.0	2.6	10.9	1.0
Disgust	9.4	**75.5**	6.3	5.7	1.6	1.6
Fear	5.7	2.6	**56.8**	15.6	11.5	7.8
Happiness	2.1	6.3	16.1	**74.0**	1.0	0.5
Sadness	12.0	0.5	7.3	2.1	**78.1**	0.0
Surprise	2.6	1.0	2.1	2.1	1.0	**91.1**

Table 9 Confusion matrix of the LDA for the full velocity fields

Input/ Output	Anger (%)	Disgust (%)	Fear (%)	Happiness (%)	Sadness (%)	Surprise (%)
Anger	**77.6**	7.8	0.5	2.1	11.5	0.5
Disgust	8.9	**77.1**	5.2	5.2	2.6	1.0
Fear	4.7	3.6	**61.5**	9.9	13.0	7.3
Happiness	3.1	6.3	14.1	**76.0**	0.0	0.5
Sadness	15.1	0.0	6.8	1.6	**76.6**	0.0
Surprise	1.6	1.6	3.6	1.0	1.6	**90.6**

The results of misclassification support the conclusion of the separability analysis conducted in the previous section. The pair of expressions with low value of separability criterion $J_2^{\mathrm{ex}_i,\mathrm{ex}_j}(x)$ are more prone to be misclassified (e.g. "fear and sadness"). The expression of fear achieves low values of separability criterion $J_2^{\mathrm{ex}_i,\mathrm{ex}_j}(x)$ for each facial expression representation and as it is expected the misclassification error is the highest. The expressions with high value of separability criterion $J_2^{\mathrm{ex}_i,\mathrm{ex}_j}(x)$ achieve high recognition rates (e.g. "happiness, or surprise").

Table 10 summarises the success rates of the recognition for the different representations included in Tables 6–9. Taking into account the *subjective* nature of the ground truth data [19], the results can be considered as reasonable.

Table 10 Summary of success rates for confusion matrix of the LDA for different representations

Feature/ Expression	Anger (%)	Disgust (%)	Fear (%)	Happiness (%)	Sadness (%)	Surprise (%)
Manually	74.5	81.8	59.9	85.4	77.6	89.6
Automatic	68.8	76.6	55.2	82.3	72.4	87.0
Deformation	74.5	75.5	56.8	74.0	78.1	91.1
Velocity	77.6	77.1	61.5	76.0	76.6	90.6

4 Conclusions

A statistical analysis of different facial expression representations based on the log-Euclidean statistics has been presented in this paper. The proposed method generates first the *mean* face by simultaneous registration of faces with neutral expression included in the training data set, thereby enabling all faces to be mapped to the common face space based on the estimated transformations. The obtained results show that the Space Shape Vectors built based on the velocity fields can be considered as an effective facial expression representation for the Statistical Shape Model. The performed tests show also that the parameterisation via stationary velocity fields in the log-domain produces slightly higher recognition rate of facial expressions than that produced by using deformation fields. The future investigation can consider extension of the proposed facial expression recognition system for the dynamic high-resolution sequences [13]. The temporal information estimated by the velocity fields can lead to improvement of the performance of the current system.

References

1. Arsigny, V., Commowick, O., Pennec, X., Ayache, N.: A Log-Euclidean framework for statistics on diffeomorphisms. In: Larsen, R., Nielsen, M., Sporring, J. (eds.) Proceedings of the 9th International Conference on Medical Image Computing and Computer-Assisted Intervention (MICCAI 2006), Copenhagen, Part I. Lecture Notes in Computer Science, **4190**, 924–931. Springer (2006). ISBN 3-540-44707-5
2. Ashburner, J.: A fast diffeomorphic image registration algorithm. NeuroImage **38**(1), 95–113 (2007)
3. Bartlett, M.S., Littlewort, G., Fasel, I., Movellan, J.R.: Real time face detection and facial expression recognition: development and applications to human computer interaction. In: Proceedings of the IEEE Conference on Computer Vision and Pattern Recognition Workshop, 2003 (CVPRW '03). doi:10.1109/CVPRW.2003.10057
4. Bishop, C.M.: Pattern Recognition and Machine Learning (Information Science and Statistics). Springer New York, Inc., Secaucus (2006)
5. Bossa, M., Hernandez, M., Olmos, S.: Contributions to 3D diffeomorphic atlas estimation: application to brain images. In: Ayache, N., Ourselin, S., Maeder, A.J. (eds.) Proceedings of the 10th International Conference on Medical Image Computing and Computer-Assisted Intervention (MICCAI 2007), Brisbane, Part I. Lecture Notes in Computer Science, **4791**, 667–674. Springer (2007). ISBN 978-3-540-75756-6

6. Cootes, T.F., Taylor, C.J., Cooper, D.H., Graham, J.: Active shape models – their training and application. Comput. Vision Image Understand. **61**, 38–59 (1995)
7. Fasel, B., Luettin, J.: Automatic facial expression analysis: A survey. Pattern Recogn. **36**(1), 259–275 (2003)
8. Geng, X., Christensen, G.E., Gu, H., Ross, T.J., Yang, Y.: Implicit reference-based group-wise image registration and its application to structural and functional MRI. Neuroimage **47**(4), 1341–1351 (2009)
9. Han, X., Hibbard, L.S., Willcut, V.: An efficient inverse-consistent diffeomorphic image registration method for prostate adaptive radiotherapy. In: Madabhushi, A., Dowling, J., Yan, P., Fenster, A., Abolmaesumi, P., Hata, N. (eds.) Prostate Cancer Imaging. Computer-Aided Diagnosis, Prognosis, and Intervention – International Workshop, Held in Conjunction with MICCAI 2010, Beijing. Lecture Notes in Computer Science, **6367**, 34–41. Springer (2010). ISBN 978-3-642-15988-6
10. Hsieh, C.K., Lai, S.H., Chen, Y.C.: An optical flow-based approach to robust face recognition under expression variations. IEEE Trans. Image Process. **19**(1), 233–240 (2010)
11. Kobayashi, H., Hara, F.: Facial interaction between animated 3D face robot and human beings. In: Proceedings of the IEEE International Systems, Man, and Cybernetics Computational Cybernetics and Simulation Conference, **4**, 3732–3737 (1997). doi:10.1109/ICSMC.1997.633250
12. Matuszewski, B.J., Quan, W., Shark, L.-K. Facial expression recognition. In: Albert, M (ed.) Biometrics – Unique and Diverse Applications in Nature, Science, and Technology. InTech, Rijeka (2011). doi:10.5772/16033, ISBN 978-953-307-187-9
13. Matuszewski, B.J., Quan, W., Shark, L.-K.: High-resolution comprehensive 3-d dynamic database for facial articulation analysis. In: Proceedings of the IEEE International Computer Vision Workshops (ICCV Workshops) Conference, pp. 2128–2135 (2011)
14. Pantic, M., Member, S., Rothkrantz, L.J.M.: Automatic analysis of facial expressions: The state of the art. IEEE Trans. Pattern Anal. Mach. Intell. **22**, 1424–1445 (2000)
15. Papiez, B.W., Matuszewski, B.J.: Direct inverse deformation field approach to pelvic-area symmetric image registration. In: Proceedings of the Conference on Medical Image Understanding and Analysis (MIUA'2011), pp. 193–197. British Machine Vision Association Press (2011)
16. Papiez, B.W., Matuszewski, B.J., Shark, L.-K., Quan, W.: Facial expression recognition using Log-Euclidean statistical shape models. In: Carmona, P.L., Sánchez, J.S., Fred, A.L.N. (eds.) Proceedings of the 1st International Conference on Pattern Recognition Applications and Methods (ICPRAM 2012), Vilamoura, Algarve, **1**, 351–359. SciTePress (2012). ISBN 978-989-8425-98-0
17. Quan, W., Matuszewski, B.J., Shark, L.-K., Ait-Boudaoud, D.: 3-D facial expression representation using B-spline statistical shape model. In: Proceeding of the Vision, Video and Graphics Workshop, Warwich. British Machine Vision Association Press (2007)
18. Quan, W., Matuszewski, B.J., Shark, L.-K., Ait-Boudaoud, D.: Low dimensional surface parameterisation with applications in biometrics. In: Proceedings of the International Conference on Medical Information Visualisation – BioMedical Visualisation, pp. 15–22. IEEE Computer Society, Silver Spring (2007)
19. Quan, W., Matuszewski, B.J., Shark, L.-K., Ait-Boudaoud, D.: Facial expression biometrics using statistical shape models. EURASIP J. Adv. Signal Process. **2009**, 15:4–15:4 (2009)
20. Quan, W., Matuszewski, B.J., Shark, L.-K.: Improved 3-D facial representation through statistical shape model. In: Proceedings of the IEEE International Conference on Image Processing (ICIP 2010), Hong Kong, pp. 2433–2436. IEEE (2010). ISBN 978-1-4244-7994-8
21. Shan, C., Gong, S., McOwan, P.W.: Robust facial expression recognition using local binary patterns. In: Proceedings of the IEEE International Conference on Image Processing (ICIP 2005), Genoa, **2**, 370–373. IEEE (2005)
22. Song, J.H., Christensen, G.E., Hawley, J.A., Wei, Y., Kuhl, J.G.: Evaluating image registration using NIREP. In: Fischer, B., Dawant, B.M., Lorenz, C. (eds.) Proceedings of the 4th International Workshop on Biomedical Image Registration (WBIR 2010), Lbeck. Lecture Notes in Computer Science, **6204**, 140–150. Springer (2010). ISBN 978-3-642-14365-6

23. Tian, Y.L., Kanade, T., Cohn, J.F.: Facial expression analysis. In: Handbook of Face Recognition. Springer, Berlin (2011)
24. Vercauteren, T., Pennec, X., Perchant, A., Ayache, N.: Symmetric log-domain diffeomorphic registration: A demons-based approach. In: Proceedings of the 11th International Conference on Medical Image Computing and Computer-Assisted Intervention – Part I. MICCAI '08, pp. 754–761 (2008)
25. Vercauteren, T., Pennec, X., Perchant, A., Ayache, N.: Diffeomorphic demons: Efficient non-parametric image registration. NeuroImage **45**(1, Supp.1), S61–S72 (2009)
26. Wang, J., Yin, L.: Static topographic modeling for facial expression recognition and analysis. Comput. Vision Image Understand. **108**(1–2), 19–34 (2007)
27. Yin, L., Wei, X., Sun, Y., Wang, J., Rosato, M.J.: A 3D facial expression database for facial behavior research. In: Seventh IEEE International Conference on Automatic Face and Gesture Recognition (FG 2006), Southampton, pp. 211–216. IEEE Computer Society (2006). ISBN 0-7695-2503-2

Printed by Publishers' Graphics LLC
AMZ20121115.11.35.26